Pan-European conference on the potential long-term ecological impact of genetically modified organisms

proceedings

Strasbourg, 24-26 November 1993

Environmental encounters, No. 20

Council of Europe Press, 1995

English version only.

Publishing and Documentation Service
Council of Europe
F-67075 Strasbourg Cedex

ISBN 92-871-2679-8

Printed in the Netherlands

Nature and environment

1. Aspects of forest management, 1968 (*out of print*)
2. Freshwater, 1968 (*out of print*)
3. Animals in danger, 1969 (*out of print*)
4. A handbook for local authorities, 1971 (*out of print*)
5. Soil conservation, 1972 (*out of print*)
6. Endangered Alpine regions and disaster prevention measures, 1974
7. Air pollution problems – Manual of experiments, 1975
8. Evolution and conservation of hedgerow landscapes in Europe, 1975
9. The integrated management of the European wildlife heritage, 1975
10. Threatened mammals in Europe, 1976 (*out of print*)
11. The effects of recreation on the ecology of natural landscapes, 1976 (*out of print*)
12. Heathlands of western Europe, 1976 (*out of print*)
13. The degradation of the Mediterranean maquis, 1977 (published jointly with Unesco)
14. List of rare, threatened and endemic plants in Europe, 1977 (*out of print*)
15. Threatened amphibians and reptiles in Europe, 1978 (*out of print*)
16. Vegetation map (scale 1:3 000 000) of the Council of Europe member states, 1979
17. Model outline environmental impact statement from the standpoint of integrated management or planning of the natural environment, 1980
18. Threatened freshwater fishes of Europe, 1980
19. European peatlands, 1980
20. Behaviour of the public in protected areas, 1981
21. Dry grasslands of Europe, 1981
22. Alluvial forests in Europe, 1981
23. Threatened Rhopalocera (butterflies) in Europe, 1981
24. Birds in need of special protection in Europe, 1981
25. Inventory and classification of marine benthic biocenoses of the Mediterranean, 1982
26. Town farms, 1982
27. List of rare, threatened and endemic plants in Europe (1982 edition), 1983
28. Nature in cities, 1982
29. The vegetation of the Alps, 1983
30. Salt marshes in Europe, 1984
31. Protected marine areas, 1985
32. European dune and shoreline vegetation, 1985
33. Ecological repercussions of constructing and using ski-runs, 1986
34. Environmental education for the farming community – Experimental schemes in Europe, 1987
35. Invertebrates in need of special protection in Europe, 1987

36. Development of flora and fauna in urban areas, 1987
37. Conservation of marine benthic biocenoses in the North Sea and the Baltic, 1987
38. The protection of dragonflies (Odonata) and their biotopes, 1988
39. Problems of soil conservation, 1988
40. Texts adopted by the Council of Europe in the field of the conservation of European wildlife and natural habitats, 1989
41. The biology, status and conservation of the monk seal (*Monachus monachus*), 1989
42. Saproxylic invertebrates and their conservation, 1989
43. Possible causes of forest decline and research programmes in Europe, 1989
44. The biological significance and conservation of Hymenoptera in Europe, 1990
45. Status, conservation needs and reintroduction of the lynx (*Lynx lynx*) in Europe, 1990
46. Conservation of threatened freshwater fish in Europe, 1990
47. Status and conservation needs of the wolf (*Canis lupus*) in the Council of Europe member states, 1990
48. Marine turtles in the Mediterranean : distribution, population status, conservation, 1990
49. Evergreen forests in the Macaronesian Region, 1990
50. Marine reserves and conservation of Mediterranean coastal habitats, 1990
51. Towards the conservation of aculeate Hymenoptera in Europe, 1991
52. The means of giving greater prominence to environmental issues in agricultural education at secondary school level, 1992
53. Revision of Appendix I (Flora), 1992
54. The wild mink (*Mustela lutreola*) in Europe, 1992
55. Status and conservation of the pardel lynx (*Lynx pardina*) in the Iberian Peninsula, 1992
56. The conservation of natural habitats outside protected areas : legal analysis, 1992
57. The conservation of European orchids, 1992
58. Balanced development of the countryside in western Europe, 1992
59. Rehabilitation of natural habitats in rural areas, 1992
60. Datasheets of flora species – Volume I, 1992
61. Datasheets of flora species – Volume II, 1992
62. Datasheets of flora species – Volume III, 1992
63. Datasheets of flora species – Volume IV, 1992
64. Threatened non-marine molluscs of Europe, 1992
65. Potential long-term ecological impact of genetically modified organisms, 1993

66. Conservation of freshwater fish in Europe, 1994
67. Status and conservation needs of the otter *(Lutra lutra)* in the western Palaearctic, 1994
68. Guidelines to be followed in the design of plant conservation or recovery plans, 1994
69. Status and conservation of the wildcat (*Felis silvestris*) in Europe and around the Mediterranean rim, 1994
70. The integrated development of the countryside in central and eastern European countries, 1994
71. European soil resources, 1995
72. Underground habitats and their protection, 1995

Environmental encounters

1. Environmental training in agricultural circles, 1987
2. Parks, plans and people – Protected areas and socio-economic development, 1987
3. Workshop on the situation and protection of ancient natural and semi-natural woodlands in Europe, 1987
4. A new management of the environment in rural areas, 1988
5. Training course for managers of protected areas in Europe and Africa, 1989
6. The situation of the brown bear (*Ursus arctos*) in Europe, 1989
7. Nature tomorrow, 1989
8. The conservation of wild progenitors of cultivated plants, 1991
9. Nature museums: tools for learning about, promoting and protecting the natural heritage of Europe, 1990
10. Colloquy on the Berne Convention invertebrates and their conservation, 1990
11. The situation, conservation needs and reintroduction of the lynx in Europe, 1992
12. The management of Mediterranean wetlands, 1992
13. The conservation of the Mediterranean monk seal, 1992
14. Wetland invertebrates, 1992
15. Seminar for managers of diploma-holding areas, 1992
16. Seminar on the biology and conservation of the wildcat (*Felis silvestris*), 1993
17. Seminar on the management of small populations of threatened mammals, 1994
18. Workshop on nature conservation in central and eastern Europe, 1994
19. Seminar on recovery plans for species of amphibians and reptiles, 1994

Planning and Management

Hedges, 1988

Farming and wildlife, 1989

Other publications

Giving nature a chance : the Bern Convention, 1984 (*out of print*)

A European cultural revolution : the Council of Europe's "Charter on invertebrates", 1986 (*out of print*)

Europe and Africa defend nature : the Bern Convention, 1986 (*out of print*)

Management of Europe's natural heritage. Twenty-five years of activity, 1987

The Bern Convention on Nature Conservation, 1991

Naturopa (periodical)

The majority of the Council of Europe's publications are available in English and French editions.

Pan European Conference on the potential long term ecological impacts of genetically modified organisms (PEC), November 24-26, 1993, Strasbourg

C O N F E R E N C E   S T A T E M E N T

The PEC was organized by the Council of Europe, with the support of the Norwegian Ministry of Environment, the Spanish Ministry of Public Works, Transport and Environment, the Netherlands Ministry of Housing, Physical Planning and Environment, and the Commission of the European Community.

One hundred and thirty participants from Austria, Belgium, Canada, Croatia, Cyprus, Denmark, France, Germany, Hungary, Ireland, Italy, Latvia, Lithuania, The Netherlands, Norway, Portugal, Russia, Slovenia Republic, Spain, Sweden, Switzerland, Turkey, Ukraine, and the United Kingdom participated in this conference. The participants were experts in a broad variety of disciplines: ecology, molecular biology, agronomy, epidemiology, population genetics and so forth.

The purpose of the Conference was to provide a scientific forum for exchange of views and information on potential long term ecological impacts of the release of genetically modified organisms, in order to review knowledge, theory and experience in this field.

Starting point of the discussions and contributions was the Council of Europe report "Potential long term ecological impacts of genetically modified organisms". In this report an approach is presented to gain insight into long term ecological effects and their underlying interactions by examining documented introductions of organisms into ecosystems.

After general introductions to biotechnology and ecology, the conference focused on several topics: insect resistant plants, virus resistant plants, baculoviruses, vaccines, fish and the potential role of monitoring. In five case studies, the purpose and technical background of a particular application was presented, followed by a presentation in which the case study was placed in a long(er) term ecological perspective. All presentations were discussed by the participants.
Issues that were discussed:

- what do we mean by 'long term' in different contexts?
- what do we mean by 'ecological impacts'?
- in assessing potential impacts, taking into account:

    * the characteristics of the organism, of the environment and the interaction between the organism and its environment;
    * the concept of 'keystone species';
    * the concept of causality;
    * differences between terrestrial and aquatic ecosystems;
    * the consequences for human health and the environment of the alternative, if a certain application is **not** carried out;

- in assessing potential impacts, distinguishing safety aspects and efficacy aspects, and distinguishing ecological and socio-economic impacts;
- the practical meaning of the 'case by case' concept;
- the potential value and limitations of monitoring and ecological generalizations;
- the concepts of direct and indirect monitoring, the value and limitations of monitoring by 'third parties' such as farmers;
- The value of predictions in relation to short term and long term impacts; how can we deal with unpredictability?

CONTENTS

WELCOMING ADDRESS

Mr. F. Albanese, Director of Environment and Local Authorities, Council of Europe

Mr. Chairman, ladies and gentlemen,

It is a pleasant duty and a great honour for me to welcome all the participants of the Pan European Conference on the Potential Long Term Ecological Impacts of the Release of Genetically Modified Organisms to the Palais de l'Europe here in Strasbourg.

Allow me to convey to you the warm welcome of our Secretary General, Mrs. Catherine Lalumière, who is unable to be with us today owing to her heavy schedule of commitments. Convinced that this conference will lead to tangible results, she has asked me to assure you of her best wishes for success in your work.

The governments of the Netherlands, Spain and Norway have given the Council of Europe valuable help in preparing this event and our heartfelt thanks are due to them.

When the Committee of Ministers of the Council of Europe added the question of genetically modified organisms to the intergovernmental programme of activities, it is true that it had no intention of tackling this issues relating to biotechnology; other organizations are active in this field. However, since the Council of Europe has been concerned, for many years, with the protection and management of the environment, especially natural habitats, its chief aim was to see to what extent our environment might possibly be damaged by the accidental or deliberate release of genetically modified organisms.

The choice of the theme of this conference is therefore a judicious one, since man may be both the architect and destroyer of his environment. This is why, in my view, the three adjectives "ecological", "potential' and "long-term" are the most meaningful in this context and discussions at the conference should focus on them as matter of priority. The Organizing Committee, to whose work I must pay tribute, was right to organize the conference around these three concepts.

You are set to tackle a most important subject, one which has aroused many hopes and fears among the public. The fears, often fomented by a press greedy for sensation, spring from the fear of irreparable damage to the environment - leaving aside the emotional reaction to some applications of biotechnology in human genetics. The hopes reside in the expectation that biotechnology will bring progress for humanity at some future time, especially in the fields of medicine, foodstuffs and clean power.

It will be your task to review the present situation, and maybe to reassure public opinion. It is vital that progress in biotechnology, and the vast possibilities opening in its wake, be achieved not only in the interest of mankind and human dignity but also without harming the interests of the conservation of nature and the environment and conservation in general.

We expect much from this conference. It must not be just "another" conference, but a forum for ideas and exchange of experience, leading to specific conclusions for the future of the natural heritage in Europe, and if possible to new action on the part of our organization.

With this wind in our sails, I wish you every success.

INTRODUCTION TO THE CONFERENCE

Mr. drs. P.J. van der Meer, Chairman of the Conference, Ministry of Housing, Physical Planning and Environment, The Netherlands

Ladies and gentlemen, distinguished participants and observers, on behalf of the Organizing Committee I welcome you to this conference. And I would like to express a special welcome to those who have come from so far.

In this introduction to the conference I intend to give you some background information on the history of this conference, place it in a broader context and finally I will focus on the purpose and the structure of this conference.

HISTORY OF THIS CONFERENCE

Let me take you back into the past and let us look at some dates and developments that preceded this conference.

1938	The term 'Molecular Biology' is used for the first time in an annual report to the Trustees of the Rockefeller foundation.
1953	DNA as a double helix; Watson and Crick.
1972	The first applications of the recombinant DNA technology are published by Jackson, Symons and Berg.
1974	The discussion in safety in biotechnology begins with the publication in Science of a letter from Berg, Baltimore and Cohen, on "The potential biohazards of recombinant DNA molecules"
1986	Publication of the OECD recombinant DNA safety considerations.
1990	Adoption of two European Community directives on genetically modified organisms.

In view of these developments the Council of Europe was faced with two possible courses of action: either to recognize the progress already made by the European Community and to extend it to the other Member States of the Council of Europe, or to seek a new approach.

For this reason the Steering Committee for the Conservation and Management of the Environment and Natural Habitats (CDPE) of the Council of Europe appointed a Group of specialists to examine the Ecological Impacts of Gene Technology.

Report on the potential long term ecological impacts of genetically modified organisms

As a start, the Group produced a report on the potential ecological impacts of genetically modified organisms. This report is included in the conference documentation.

Chapter 2 of that report gives a general survey of potential impacts of the use of genetic modification, both positive and negative. It reaches the following

conclusions:

- Biotechnology can contribute to sustainable development.
- In order benefit maximally of the potentials biotechnology, consideration of possible unintended side effects is imperative.

In Chapter 3 and Chapter 4, the Group summarized, on the basis of the survey, some generally accepted principles and approaches for safety mechanisms, risk assessment and risk management.

To start with, the Group recognized that the system presented in the OECD Blue Book and in the EC Directives offers a sound basis for adequate risk assessment and risk management.

The following conclusions can be drawn in relation to risk assessment and risk management:

- Safety in biotechnology is achieved by the appropriate application of risk assessment and risk management, which include the following steps:
  a. Supply of information.
  b. Review prior to experimental release
  c. Application of safety conditions
  d. Monitoring during and after the experiment
- Risk assessment entails a primary focus is on the characteristics of the organisms, including a focus on GMOs.
- A general principle for safety is 'step by step'.
- Contained use: classification of risk management categories.
- Releases: initial case by case review.

Chapter 5 of the report presents an approach, which focuses on the long term effects. This approach is meant to provide additional information to the current risk assessment programmes; it is not meant in any way to replace the current risk assessment approaches. An attempt is made to gain insight into ecological effects and their underlying interactions by examining the introduction of organisms into ecosystems. The approach presented intends to use the wealth of existing information and knowledge on biological invasions, epidemiology and so forth.

A strategy for further work is presented, and is illustrated by several case studies in which information was extracted from introductions of non genetically modified organisms.

Risk assessment consists of an examination of possible effects (hazards, harm), and an assessment of the probability that these effects will occur. There are a number of effects that can be considered :

- the very first effect of a new gene in an organism, the production of a gene product.
- the effect of the gene product on an organism.
- the short term effect of the organism with the new trait on the ecosystem where it is used
- the long term effect of an organism which influences certain ecosystems because of its excessive growth.
- the effect of the new gene of transferred DNA to other organisms, etc.

effect 1 effect 2 effect 3 ....?........................> LONG TERM EFFECT

------------>------------>------------>------------>------------>

In addition to these steps in current risk assessment, the report intends to place an emphasis on **long term** effects. Attention is given to a description of potential long term effects, because by a description of the kind of long term effects that can be expected is it possible to judge whether or not certain risks are acceptable. Furthermore, by a giving a description of the kind of effects that may be envisaged, adequate monitoring can be developed.

In order to identify the mechanisms through which effects may occur, the report suggests that several steps have to be taken:

a. identification of an effect.
b. characterization of the ecological interactions that caused the effect.
c. characterization of the phenotypic or physiological traits of the organisms which led to the interactions.
d. characterization of the genetic traits that lead to phenotypic or physiological traits of the organisms.

```
effect 1    effect 2    effect 3    ......?........................>        LONG

------------>------------>------------>------------>------------>        TERM

...............<.......?........<...---------<------------<------------   EFFECT

            step d      step c      step b      step a
```

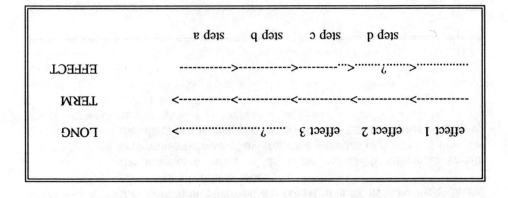

Several case studies in chapter 5 serve as the basis for the identification of long term effects of the introduction of organisms. The identified effects are:

- spread of diseases;
- reduction of population numbers;
- occurrence of pests;
- hindering growth of other organisms;
- contamination of the environment by toxic chemicals or hazardous substances;
- changes in geochemical and physical processes;
- loss of beneficial characteristics or resources;
- loss of genetic diversity.

Subsequently, the characterized the ecological interactions that caused the effect as:

- pathogen - host interactions;
- predator - prey interactions;
- competition;
- mutualism / commensalism;
- allelopathy;
- interactions with the abiotic environment;
- gene flow within and between populations.

PURPOSE OF THE CONFERENCE

This brings me to the purpose of this conference. The approach suggested in the report was applied in some twenty case studies, carried out by three ecologists of the Group.

On the basis of the results of the case studies, the Group decided that the time had come to present this approach to a forum of experts, such as yourselves. As a try out, a seminar on this topic was organized in 1991, in the Netherlands, by the Netherlands Ecological Society. Strengthened by the constructive dialogue in that seminar, the Group decided to organize a Pan European Conference on this issue. Soon after, an Organizing Committee was formed, which immediately started the preparations. The Council of Europe and the Governments of Norway, Spain and the Netherlands financial supported this conference.

The purpose of this conference is threefold:

1) stimulating the dialogue between molecular biologists and ecologists

2) drawing the attention to potential long term ecological effects;

3) peer review: asking the opinion of international experts on the usefulness of the approach suggested in the report.

STRUCTURE OF THE CONFERENCE

First of all, I should point out that the Organizing Committee deliberately decided to have plenary sessions only. This is done to have all the necessary expertise available during all discussions.

To start with, prof. Schell will give us an update of the biotechnological developments and dr. Crawley will set the scene for ecological theory.

Following these two general sessions, 5 topics will be presented, all in the same format: first a rapporteur will shortly set the scene for that topic, followed by a presentation on a particular biotechnology case and finally followed a presentation in which that case study is placed in a broader ecological, long term perspective. After each session there is ample time for discussion, chaired by the rapporteurs.

Realizing that some of the issues do require a more detailed discussion in smaller groups, we decided to make one exception on the plenary sessions: on Thursday evening each of the five topics will be discussed over dinner in a separate group.

Ladies and gentlemen, the Organizing Committee would like to emphasize one of the aims of this conference: dialogue between scientists of different disciplines.
We therefore call upon you to really enter this dialogue, to share with us your expertise.

We call upon you as ecologists, molecular biologists, plant breeders, virologists and so forth, to teach us from your knowledge of what happened in the past in order to learn about the future.
We call upon you to assist us in separating scientific facts from science fiction.

Having expressed this unanimous wish of all members of the Organizing Committee, I wish you a fruitful conference and thank you for your attention.

SETTING THE SCENE FOR BIOTECHNOLOGY

PLANT BIOTECHNOLOGY: STATE OF THE ART IN DEVELOPED COUNTRIES AND RELEVANT SAFETY CONSIDERATIONS

Prof. dr. J.S. Schell, Max-Planck-Institute for Plant Breeding Research, Germany

SUMMARY

Agricultural productivity is essential in the face of an increasing world population and limited cultivatable land. Plant breeding is one of the few methods available, and one of the most effective, to maintain and improve productivity without destroying the environment. Gene technology is an essential element in modern plant breeding. Examples of transgenic plants resistant to herbicides, viruses, insects and fungi, and applications related to food quality and the production of valuable chemicals will be presented. Because plant genetic engineering is based on the transfer of defined genes, the risks are predictable and can therefore be controlled. Over 1000 field tests of transgenic plants have already been carried out without indication of dangers or unpredicted consequences. Appropriate regulations should therefore be able to ensure both the safe and efficient use of this essential technology.

INTRODUCTION

Gene technology as a new and rapidly developing technology had its beginnings in the late nineteen-seventies. In the developed world, most especially in Germany, environmental organizations, political parties, the media, religious and secular groups and the consumer reacted negatively and with a very restrictive attitude toward this technology.

However, a society which is responsible, which concerns itself not only with the wellbeing of its present citizens but also with future generations and which considers environmental protection seriously cannot afford to ignore or hinder progress in plant biotechnology. Even in the near future, agriculture, whether global or regional, whether intensive or extensive, whether industrial or familial, will no longer be optimally productive without an important contribution of new scientific knowledge and without responsible application of the best and most effective technologies. This is particularly true for countries in the developing world. Regulations should therefore primarily focus on making the use of this essential technology safe and efficient, rather than concentrating on hypothetical and often unrealistic dangers.

To evaluate the possibilities of plant biotechnology one should keep in mind the following:

1. Agriculture, as it is practised currently, is one of the biggest sources of environmental pollution. Continuation of these practices can lead to rapid and possibly irreversible deterioration of the environment, thus putting the sustainability of agriculture in question.

2. Agriculture must be productive in order to be commercially viable, and socially and environmentally acceptable. If agriculture is to remain an attractive occupation, it must be economically rewarding. If one wants to diminish the negative impact of agriculture on the environment, one should optimize productivity, i.e. maximum quality and yield for a given input, such that one can reduce input and at the same time conserve or even improve quality and yield.
3. Plant breeding is one of the few and one of the most effective methods, to improve agricultural productivity without simultaneously destroying the environment. This is true for the industrialized world, perhaps even more true for the developing world and holds for both intensive and extensive agriculture.
4. If plant breeding is to contribute to the solution of the enormous problems which we must face in the next decades, then the best techniques must be used including genetic engineering. The resulting plants must then be compared to already available crops for their effect on health and on the environment.

In relation to gene technology in particular one must remember:

1. Transgenic plants and microorganisms can help to diminish the negative environmental effects of intensive agriculture.
2. More than 1000 tests in the field have already been carried out around the world and have given no indication of real dangers or consequences which had not been predicted.
3. Transfer of genes occurs in nature (as in the case of *Agrobacterium tumefaciens*).

STATE OF THE ART

The basic underlying fact is that the genetic code is universal. Therefore, it is possible to express any genetic trait of any organism in plants. In practice this allows us to make plants expressing a defined property of other types of organisms, be it bacteria or yeast or animal cells. This vastly increases the potential for plant breeding.

Two scientific breakthroughs underlay genetic engineering in plants. First was the development of recombinant DNA technology which made it possible to isolate individual genes from any organism. The second was the discovery in the 1970s that there are bacteria in the soil which transfer genes into plants, i.e. *Agrobacterium tumefaciens.* In nature this bacterium is a pathogen which causes tumours in plants. It was discovered that these tumours are the consequence of the transfer of a sequence of DNA, called transferred DNA (T-DNA), from the bacterium into the nuclear genome of the plant cell. The transformed plant cells grow as tumours because the T-DNA contains oncogenes. If one removes the oncogenes from the

T-DNA, the system is still capable of transferring DNA into the plant cell, and one can therefore use T-DNA based gene vectors to introduce foreign genes into plants and obtain so-called transgenic plants.

If one wants to express a foreign gene in plants, it is usually not effective to transfer the whole DNA sequence, the whole gene. Although the coding sequence is universal and should work in every cellular environment, it was soon found that a coding sequence must be surrounded by plant-specific promoter sequences in order to be functional in plants. If, therefore, one wants to express a bacterial gene in plants, one has to make a chimeric gene combining the coding sequence of the bacterial gene with the promoter sequence from a known plant gene. It is possible to predetermine in which organs of the plant a chimeric gene will work by using a promoter from a plant gene which is specifically expressed in the desired organ. In other words, it is now possible to tailor-make genes of any origin so that they function in a predetermined organ of the plant.

The first transgenic plants were obtained in 1983. These plants expressed chimeric genes that protected them against toxic chemicals such as antibiotics. The first were resistant to kanamycin, because they expressed a bacterial gene producing an enzyme (neomycin phosphotransferase) able to detoxify kanamycin. This gene has proven useful in the isolation of transgenic plants. Many plant cells have the remarkable property of "totipotency"; vegetative tissues in tissue culture can be induced to regenerate fully fertile plants. If such a regeneration, e.g. from a leaf derived tissue, is performed in the presence of a toxic chemical like kanamycin, only cells that can resist kanamycin will be able to produce plantlets. Therefore, transgenic plants can be selected for by this method and any foreign gene can be introduced linked to such a selectable marker gene. In tobacco, the whole process from leaf to fertile transgenic plants, takes only about 3 months.

More recently other methods have been developed for gene transfer into plants. One of them, developed in the U.S., is the particle gun in which a cartridge with small metal particles soaked in a DNA solution is used. The particles with the DNA are shot through the tissue of the plant and some of the DNA remains in the nucleus and can be expressed. Several other methods such as electroporation or direct uptake by protoplasts have also been used.

Thus far most transgenic plants have been obtained using the natural *Agrobacterium* Ti-plasmid T-DNA vectors. The list of transformed plants is getting longer every day and includes crops such as rice and corn. It can be safely predicted that provided sufficient work is invested, any type of plant, every crop or vegetable can be transformed with one or another of these methods. With regard to tropical crops, with largely regional value, it will not be possible to leave it to multinational industries to provide the technical know-how. Publicly funded, national and international agencies will have to assume this responsibility.

Historically, the first genes to be developed for molecular plant breeding were herbicide resistance genes. Let us consider a herbicide such as bialaphos, which is well known in Japan. In Europe a similar herbicide, phosphinotricin with the commercial name "Basta", is chemically produced. A bacterial enzyme (acetyltransferase) can modify this herbicide into an acetylated form which is no longer toxic. By introducing this bacterial phosphinotricin acetyltransferase into various plants, the resulting transgenic plants were made selectively resistant to the herbicide. Such transgenic tobacco plants were used in the first ever field experiments in 1987 in France.

The ideal modern-day herbicide would be one which is non-toxic to animals and humans, not harmful to the environment and selectively toxic for weeds as opposed to agriculturally important crops. Bialaphos and phosphinotricin are non-toxic to animals and humans and easily degraded in the soil so that they disappear after their usefulness in controlling weeds is over. Thus, they approach the above ideal. However, their weakness is that they are not selective, i.e. they have not been developed to be used in conjunction with agriculturally important crops, and kill the crops as well as the weeds.

A small digression: It is very unfortunate that in the debate on plant genetic engineering these non-selective herbicides have often been referred to as "total herbicides", which immediately suggests to everyone, even to fairly informed people, that a total herbicide is a general poison, which is not the case. It is better to be more specific and correct and call them "non-selective".

"Selective" herbicides have been designed. This is possible because some crop plants have genetically determined mechanisms to detoxify these particular "selective" herbicides. The important question was whether one can find genes that produce enzymes capable of detoxifying "non-selective" herbicides.

Taking again the example of phosphinotricin: Phosphinotricin, an analogue of glutamic acid, is a potent competitive inhibitor of the enzyme glutamine synthase. Its use leads to the accumulation of ammonia in plant cells which is toxic and kills the weeds but also most crops. The natural product bialaphos is very similar to phosphinotricin (identical with the exception that it is a tripeptide rather than a single amino acid) and is produced by some soil microorganisms. Microorganisms use such products to compete with other microorganisms. The bacterium *Streptomyces hygroscopicus* which produces this tripeptide, protects itself by detoxifying the product via acetylation. The acetylated product is no longer a competitive inhibitor because it cannot bind to the enzyme.

Thus, in order to protect plants specifically against phosphinotricin, the strategy was very straightforward. From the streptomyces strain that produces the detoxifying acetylase enzyme, one can isolate the corresponding gene and then modify this gene so that it will work in plants. This turned out to be relatively simple and has been done by several groups. The results were as expected. The growth of transgenic plants expressing the transacetylase gene cannot be distinguished, even in the presence of the herbicide, from the control normal plants grown in the absence of the herbicide. Thus, in this case, the goal of breeding by the addition of a single defined gene was achieved: genetic information, originating not from another plant but from soil microorganisms, was used in order to contribute to an important agricultural goal.

Let us consider the rate at which this technology is being developed. The first transgenic model plants (kanamycin resistant tobacco) were obtained and tested in the laboratory in 1983. In 1987 the first herbicide resistant tobaccos were field tested in France. It is planned that around 1995 a number of crops (e.g. soybean, rape seed and perhaps corn and rice) that are resistant to highly effective, easily degradable herbicides, will be commercially available. Another transgenic crop that is practically ready for commercialization by a U.S. Company is insect-tolerant cotton.

Let us take this example to illustrate pest control by plant biotechnology. The flowers of cotton are attacked by the bollworm (*Heliothis zea*) which results in dramatic yield losses. Some bacteria (*Bacillus thuringiensis* or B.t.) produce peptide-toxins which bind to specific receptors in the intestine of certain insects and are thus specifically toxic to them. By taking the genes that code for these peptide toxins and expressing them in plants, several groups have been able to produce transgenic tobaccos, tomatoes, potatoes, cotton, corn etc., which express the bacterial toxin in their leaves, flowers or other organs and are thereby effectively protected against attack by some insects. The insect larvae will feed from these plants, but as a result of the presence of the toxin their digestion is severely impaired, they stop feeding and ultimately die.

The results of various field experiments indicate that a significant measure of additional crop protection can be achieved by these genetic means. The combination of growing plants harbouring different genes which reduce their sensitivity to pests, with responsible use of environmentally acceptable pesticides, can be expected to provide great advantages both in making agriculture more efficient and economical and in reducing its negative impact on the environment. Indeed, by combining several different methods to achieve pest control, one will drastically reduce the probability that any of these methods will become obsolete as a result of adaptation by the pest.

Thus, useful pesticides would not rapidly become inefficient and could be used at lower concentrations and the biological control achieved by the products of a number of specific genes would not as readily be overcome by the pests. In this respect it may be of importance to mention that various other strategies aimed at the genetic containment of deleterious insect populations are being investigated. The goal would be to combine for one and the same crop several different insect controlling properties. Thus, one could keep insect populations in check and by appropriate regulations and agronomical procedures one could prevent the total eradication of certain insect populations. The aim cannot be and does not have to be insect eradication but instead control over the size of insect populations. It is to be expected that adequate regulations will require an integrated pest management approach and companies involved in the marketing of insect tolerant crops are developing effective integrated pest management schemes. Virus tolerance is another example of successful plant biotechnology.

Let me now discuss some recent work which is still in an earlier phase of development. Fungi produce some of the major diseases in plants. Some, but not all fungi, have a layer of chitin at the surface of their growing tips (at the growing end of hyphae). At the top of the hyphae the chitin layer is not protected by a covering layer of polyglucans and mannans. So, this part of the fungus has an exposed chitin layer. With the help of the scanning electronmicroscope Israelian researchers showed that the growing tip of such fungi can be damaged by chitinases (i.e. enzymes that degrade the chitin polymer). By reducing the growth rate of fungal hyphae in plants the normal mechanisms by which plants protect themselves against fungal attack could possibly control the fungus. By expressing a bacterial chitinase gene in plants one could hope to reduce the rate of growth of fungal hyphae and thus allow the plants to protect themselves against these pathogens. This approach was tried in our laboratory by my collaborator J. Logemann in collaboration with a group in Israel (Prof. I. Chet) who isolated a chitinase gene from bacteria capable of stopping the growth of some of the fungi. A chitinase gene (ChiA) was expressed in tobacco seedlings, and whereas in a control many of the seedlings could not grow in soil heavily infected with spores from the Basidiomycete *Rhizoctonia solani*, the transgenic seedlings appeared to be largely protected against this pathogen.

Our search for other genes producing fungicidal proteins led us to study the RIP gene from barley. This gene, which is normally expressed in barley seeds, codes for a ribosome inhibiting protein (RIP). This protein inactivates ribosomes by N-glucosidase modification of some of the nucleotides in the 28S subunit RNA of the ribosomes. Interestingly this ribosome inhibiting protein does not inactivate the ribosomes of plants, but it will inactivate the ribosomes of a number of fungi. In barley this protein is only made in the endosperm of the seeds and apparently protects seeds against fungal attack. The barley RIP gene was cloned by a group at the Carlsberg Laboratory in Denmark. This gene was modified in our laboratory and introduced into tobacco. RIP transgenic tobacco was shown to resist attack by *R. solani*. A next step will be to combine RIP and chitinase in the same plant with the hope to get even better protection. Other research groups around the world have developed other "antifungal" genes and are presently testing the effectiveness of various combinations of these genes.

Genetic engineering has thus far not only produced crops with improved protection against various biotic stresses but also some with improved food quality. The most advanced example of this type of application is presently discussed very hotly in the U.S. and concerns tomatoes with an improved ripening control. It is well known that ripening in tomato and other vegetables and also in fruits such as papaya is controlled by a simple chemical signal, i.e. ethylene. The biosynthetic pathway of ethylene in plants is well known. The direct precursor of ethylene is aminocyclopropane-1-carboxylic acid (ACC).

There are two ways to slow down ripening by slowing down the formation of ethylene. One way, which was successfully used by D. Grierson in the U.K. and which is developed by I.C.I., is the expression of a so-called "antisense gene" in transgenic tomatoes. An antisense gene of the ethylene synthase was thus used successfully.

Another technology was recently developed at Monsanto. Some bacteria can grow on ACC as sole nitrogen source by deaminating ACC and changing ACC into alpha-ketoglutarate. The enzyme used in this reaction is called ACC deaminase. By expressing the bacterial ACC deaminase in plants it is apparently possible to reduce the level of the ethylene precursor ACC and thus to reduce ethylene synthesis. This in turn resulted in a marked slowing down of the overripening of tomato.

By slowing down and controlling ripening it ought to be possible not only to increase the "shelf life" of some vegetables and fruits but also to improve their taste because taste develops relatively late in development. The FDA agency in the U.S. has recently proposed some rules to regulate the commercial release of a product which will go into the food chain. The rule sensibly stipulates that it is not the way the food product was produced that is of importance. What one has to test are the new properties of the food product.

In the same line of thinking, significant progress was recently made in attempts to use plants for the production, not of food or ornamentals, but of valuable renewable chemical feed-stocks. Two of the most important sources are carbohydrates and lipids. Results obtained thus far show that it will be possible to use plants to produce tailor-made starches or lipids with fatty acids with predetermined chain lengths. Work along these lines has progressed considerably not only in the U.S. but also in Europe. What one would like to do, for instance, is to increase the level of starch in tubers of potato from 20 to 23%, e.g. by expressing bacterial genes involved in starch biosynthesis. What would also be of commercial importance is to change the ratio of amylose to amylopectin, to make potatoes producing either only amylose or only amylopectin.

These examples show that present day biotechnology has the potential to improve agricultural productivity (via better disease and pest control) while reducing its negative impact on the environment (e.g. by the drastically reduced use of environmentally safer agrochemicals). However, this is only the beginning. The main benefit to be expected from the use of gene biotechnology is a very rapid and indeed dramatic increase in our knowledge of the mechanisms that control plant growth and yield in reaction to various biotic and abiotic (climate, soil composition etc.) stresses. Biotechnology also has the potential to make a number of plants, which presently have no agricultural value, into important "cash-crops". Indeed, many plants are known to produce highly valuable chemicals (e.g. pharmaceuticals). However, the plant genes responsible for the control of the rate of synthesis of these valuable chemicals are usually active at a low level and only under some circumstances. By the technique of "gene activation" via the random insertion of so-called promoter elements derived from very active plant genes, it is possible to activate such "lazy" genes and thereby significantly increase the production of valuable chemicals from a variety of plants that presently have no agricultural significance.

SAFETY CONSIDERATIONS

The risks of the application of gene technology to plant breeding are relatively small. Most importantly, they are predictable and therefore controllable. This is because one is dealing with defined genetic changes and one can avoid any risky changes from the start. In addition, the genetically modified plant will often, although not always, be replanted in its original environment.

Genetically modified plants or their seeds should only be given to the breeder or farmer when they have been thoroughly tested under exactly defined conditions which will vary from case to case. The design of these tests should be based on:

1. The nature of the transferred gene
2. The surroundings of the planted field
3. The existence of related plants in the surroundings.

Herbicide-resistant plants should be tested together with non-resistant plants under the same selective pressures. In the case of insect resistance it is important to measure possible effects on other insects. All field tests must include determination of environmental impact.

The organizers have provided the speakers with the following scheme. One way to think about risk assessment is to think about how much DNA is being introduced. Personally, I think this is useful. In conventional breeding one crosses plants that are related. As a consequence two different total genomes are mixed, literally hundreds of thousands of genes and enormous masses of DNA. Breeders then carefully select from the offspring that provide the plant with interesting new characteristics, interesting for agriculture, for human health and for the enjoyment of food. Thus, in conventional breeding the amount of new DNA added can be considerable. But so is the experience that has been gained by traditional breeding, an important consideration for safety and health. Until now, ecological effects have often not been of concern to breeders.

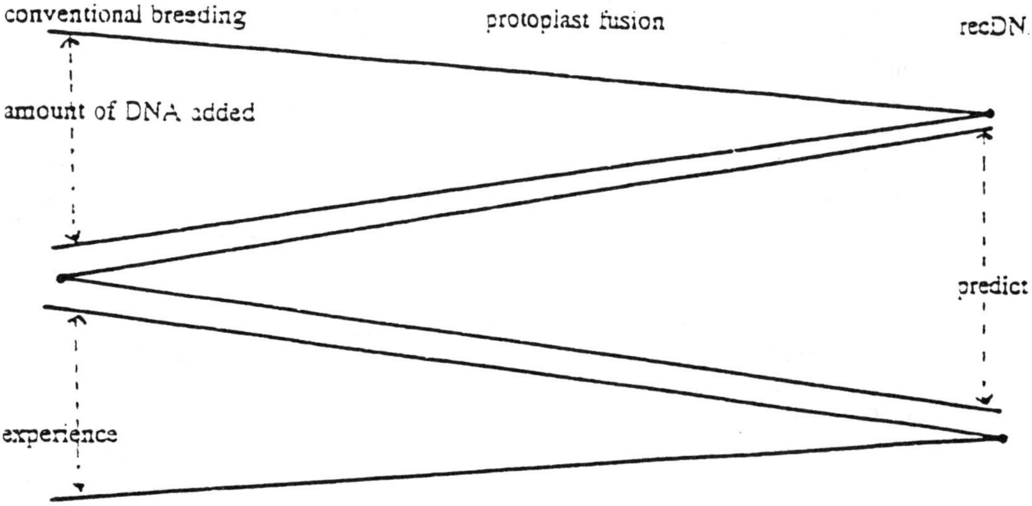

Referring again to the scheme: The less DNA you add, but even more importantly, the better one understands the function of the DNA that is being introduced, the better one will be able to predict the consequences. As has been said before, the introduction of a new gene does not necessarily change the basic biological and physiological properties of the crop in which it occurs.

This being said however, if a metabolic gene like in the case of a herbicide resistance gene, one should certainly worry about the metabolic products. For example, in the case of the acetylated herbicide accumulating in a plant, one should worry about which chemicals are made by degradation of the acetylated form and what the physiological properties of these new products are. But if the function of the introduced genes is not known, as is often the case in conventional breeding, one cannot predict the consequences. What breeders have done is to test new varieties over many years before commercialization. Thus, it is not correct that the technology *per se* is the danger. It is important to realize that molecular breeding will also be carried out by breeders with all their knowledge and experience, not by molecular biologists in their laboratories. Therefore, the high predictability of recombinant DNA will be combined with a high level of experience. All the experience from conventional breeding is relevant to plants made with recombinant DNA.

Work done by colleagues at KeyGene in Wageningen, The Netherlands, will serve as an illustration. Using a technique called AFLP (Amplified Fragment Length Polymorphism), they analyzed a number of tomato varieties that are tolerant to nematodes. Traditionally breeding, using a wide cross with wild species, produced tomatoes with natural resistance. The amount of foreign DNA from the wild species, as determined by AFLP, was 15 megabases which contains not only the genes attacking the nematodes, but also many unknown genes. Then, by improved breeding based on map positions, the amount of foreign DNA was reduced to 7 megabases of which only one gene was of specific functional importance. Finally, due to new technical developments, the amount of foreign DNA was further diminished to 500 kilobases. Compared to this, recombinant DNA can introduce a single or a few genes. So, in other words, this theoretical scheme proposed by the organizers, is indeed relevant because we know that traditional breeding leaves different amounts of unknown DNA behind in the plant. Plant molecular biology is developing methods whereby one know how much DNA is being added, and the functional characteristics of this DNA.

The days when it was said that one needs to worry only about transgenic plants are over. It would be wise to be concerned about all agricultural practices and all forms of breeding. The goal of this meeting is to work realistically. This is important, and I would warn that by taking an overly restrictive attitude, a number of consequences to the environment could result. First of all, at the very least, the development of biotechnology as a major tool to improve the environmental impact of agriculture will be impaired. Agriculture is one of the main polluters of our world and yet we cannot do without it. So, unless one explores and uses the possibilities of plant biotechnology, one will miss a major opportunity to improve the situation by breeding. Second, the potential of biotechnology is only now beginning to be realized. Research and development in this area will certainly continue because of its economic value and because of the need of the world to have agriculture. Without intelligent input, development is likely to go in directions that do not consider environmental impact. If one wants to improve the environmental impact, one has to regulate in a way that will be useful for the environment. The danger of overregulation is to prevent development and therefore to preclude the capacity to help in some or all parts of the world.

CONCLUSION

Plant biotechnology can significantly help to make intensive agriculture less damaging to the environment and to make low-input (organic) agriculture more productive. It should be possible to improve major crops (rice, wheat, corn, soybean, potato, etc.) as well as more regional crops.

If used wisely and responsibly, there is no inherent danger in these new methods. It would be shortsighted and irresponsible not to make optimal use of these methods. They can relieve at least some of the tensions that already exist and will dramatically increase in the near future because of population growth and the intensive or unproductive use of land for agricultural production. These technologies are also very likely to help increase the economic value of agriculture both in the developed and in the developing world.

PLENARY DISCUSSION

Question:
I want to complement my colleague for his enthusiastic speech. In general, I support his views. Nevertheless, I want to raise out some points which came to my mind when I listened to him. One point is the Monsanto example which he gave about a Bt treated cotton. I think that the Monsanto efforts are quite interesting but at the same time when Monsanto provides IPM receipts around this Bt modified cotton, then I think it is a little bit naive to believe that such recommendations are going to be followed. We know that similar recommendations have been given by ordinary chemical companies selling pesticides: They are not followed. So, though I like the example, we have to be very careful. Companies want to sell and they are not really interested in the ecological consequences. The ecological consequences of Bt plants are not yet known and we may place some question marks there. Another point I would like to make is this: Professor Schell has indicated that *Agrobacterium* effects in plants are normal. This is true. Many plants suffer from *Agrobacterium* induced tumours. *Agrobacterium* is just a simple plant disease organism. However, if he wanted to point out that these changes induced by *Agrobacterium* become inheritable, then I place a question mark there. A recent literature survey which went up to the 1st January 1993 has given only one minor indication that in nature there is horizontal gene transfer. It is only one minor indication in one tobacco species and that is all we could find. The literature does not indicate any mechanism that could explain how horizontal gene transfer between plants or between other organisms and plants could happen in nature. So may be there is in the enthusiasm a little bit of exaggeration.

Response:
My colleague has made three points and I agree with all three. The first point is that enthusiasm can indeed carry you further away than may be is intended. I think an enthusiastic talk may be both more and less didactic than it is aimed to be. Secondly, if I have given the impression that because Monsanto makes these slides saying that they are going to recommend Integrated Pest Management, that they will also enforce this in practice, then my enthusiasm definitely carried me too far. I hope I said, and if not, I am glad to have the opportunity to state it now, that what I see as the responsibility of groups like this one is making sure that, through regulations, this good advice will actually be implemented. I would say that this is a good case for regulations. The third point concerns *Agrobacterium*: I hope that I did not say that *Agrobacterium* effects are normal. I said they are natural. And I want to emphasize the point that the naive illusion of many people, propagated through the media, that anything natural is okay is dead wrong. Nature can be extremely dangerous. *Agrobacterium* causes a disease. This is "natural". What I attempted to illustrate, and may again have covered up with too much enthusiasm, is that this mechanism of introducing foreign genes may have contributed to evolution. When, in nature, transfer involves the whole T-DNA, tumours are formed. However, incomplete T-DNA transfers, which also occur naturally, can lead to plants with new growth properties. If these properties contribute to invasiveness, *Agrobacterium* gene

transfer can be expected to have played a role in the formation of new plants. All I want to say is that gene transfer is natural. That does not mean it is good or bad. I want to make the point that not everything that is natural is okay. The argument of naturalness cannot be used. So, thank you for raising these points with which I agree.

Question:
I would like to discuss two examples you have given us. First, the case of herbicide-resistant tobacco: Do you consider the increased use of herbicide in the future a long term effect? Second, about the chitinase-enzyme: Will this have an impact on non-pest mould?

Response:
In regard to the increased use of herbicides: Let me repeat that it is important to know which herbicide one is talking about. Biotechnology will hopefully lead to an increased use of herbicides that are degradable and that have better environmental properties. I think that the expectation is that their use will be increased. Depending on how it is done, this can lead to a decrease in the total amount of chemicals used for weed control. The current slogan in research is "One gram per hectare and bio-degradable". It is not yet realized and I do not know whether it will be achieved But certainly, the consequences could either be an increased use of herbicides in general, when purely economical parameters prevail, or it could be an increased use of environmentally safer herbicides, or a decreased use of herbicides because of the increasing use of highly powerful and environmentally safer herbicides. The outcome will depend on economical, agronomical, social (how much need for food there is) and regulatory developments.

In regard to chitinase, I should remind you that most plants have chitinases. In fact, I do not think I go wrong by saying that there are no plants without chitinases. It is important to realize that most plants survive most fungal diseases; they are only affected by a few. During evolution some fungi have managed to adapt themselves to the production of chitinases by their host plants and have become pathogens. What we attempted to do is to shortcut evolution by challenging the pathogen with the production of chitinase of a quality and in ways to which the pathogen is not adapted, e.g. with a chitinase that will be expressed continuously and therefore inhibits the growth of the fungus faster than the normal chitinases of the plants. What would not be correct to say is that the total amount of chitinase activity that fungi will be faced with is going to be dramatically different, because in nature chitinase is one of the mechanisms preferred by plants as defense against fungal diseases.

Question:
You said inserting one, single gene into a plant does not change the characteristics of the plant; it is still tobacco or a tomato. I generally fully agree with that. As a competent authority I am more and more confronted with all kinds of notifications of plants not only containing one herbicide resistance gene or a chitinase gene but containing a long list of genes in different kinds of combinations. Could you give us

any idea when you will leave your statement with regard to the number or the kind of genes, if they start to be inserted in combinations?

Response:
Again, this question allows me to come back to a point in this schema: The fundamental problem is not the amount of DNA or the number of genes, but rather how much one understands their function. Therefore, I do not think there should be an artificial limit on how many genes may be introduced. One should ask how many unknown gens have been introduced along with known ones. Unknown means ground for questioning, although it does not necessarily mean danger. Therefore, I do not see any reason for a regulator to say there should be a limit of ten or fifteen genes. If one now says fifteen, probably nobody will object because it is unlikely that one will be able to insert more in the near future. The crucial point is what one knows about the introduced genes. The kind of gene one introduces is of importance. In the case of antisense genes, for instance, one would have to know how many targets these genes have. It is true that the more genes, the more difficult the consequences are to predict because of the potential of gene products to interact. In this context, I would go back to the proven methods of breeders, namely, that before large quantities of a new breed are released, they are tested for consequences and interactions. So, in summary, I think it is not the number of genes but rather their quality that counts. The case-by-case approach seems to me the only way to proceed.

Question:
Would you agree with me that this statement is limited to plants, whereas microorganisms can be totally changed by giving them one other plasmid or a few other genes?

Response:
I certainly agree with you that what I said was limited to plants. May be I should have stated that at the beginning. However, since I was, once upon a time, a microbiologist, and *Agrobacterium* is a bacterium, I wonder whether you are right in saying that the situation with microorganisms is totally different. When you say a plasmid, which is usually not one gene (I do not know a plasmid consisting of one only gene), I would say it is probably true for microorganisms as well. If one knows all the genes introduced and their functions, the degree of predictability increases significantly. So, I think it is again a case of how much you know, rather than what organism.

Question:
We are setting the scene here as far as agriculture is concerned. And I think most of people here are from countries with overproduction. We have to remember that at least, if not more, two thirds of the world population is in a situation of equilibrium or tending to underproduction and that the curves between productivity and population increase are beginning to move apart very, very rapidly. So, I think that in considering the long term effects of biotechnology we also have to consider the long term effects on human beings; I mean we are part of the ecosystem. I was

wondering whether professor Schell would like to comment on some of the possible impacts that biotechnology could have in developing countries.

Intervention:
I would like to remind you that this is a conference limited to the long term ecological impacts of the applications of this technology. However interesting socio-economical or ethical impacts on human beings in terms of food supply might be, I would like to limit ourselves as scientists to this particular part. It is very tempting to go beyond that border, but I have seen too many conferences just floating away, simply because we allow ourselves to go into too many different directions.

Response:
I would have escaped going wrong anyway because I was planning to answer that question from an ecological point of view. I come back to one of the things I said: If we avoid nuclear war and other disasters, one of the major impacts on our environment will be caused by agriculture. I only have to think about the "slash and burn-agriculture" which people resort to in some regions when they have no other way to survive the next season. So, if we do not solve the problem of productivity, the developing world will face environmental degradation caused by the very essence of human nature. Hungry people cannot be asked either to be very generous or highly responsible. They are going to do what farmers traditionally do when they have to choose between dying and eating the seeds for the next crop: They eat the seeds. If you have the choice between dying and destroying your environment, I am afraid you destroy your environment. So, I think, without wanting to become too emotional or over-dramatic, if we do not solve the productivity problems, there will be a major impact on the environment. It is therefore appropriate to try to influence the developments in this area in such a way that will allow increases in productivity without destroying the environment, or even, when one is very optimistic, with improving the environmental impact. In previous days, productivity increases were often associated with environmental degradation.

Question:
I would, if I might pick up from the scheme presented by professor Schell, like to notice the similarity of this concept of increasing the predictability to the concept of familiarity that has been developed in more recent OECD work. The chairman was kind enough to refer to the 1986 OECD Blue Book during his presentation but the OECD is not a museum for the conservation of its past triumphs. In January 1994, the OECD will be publishing three new publications; on the scientific considerations for safety of scale up of trials with genetically modified plants, a historical review of crop breeding practices and a report on the analysis of the field releases. And finally, the OECD will provide the conference participants with the Preamble document outlining some of these newer concepts.

Response:
I must admit that when I borrowed this scheme, I thought experience and familiarity were synonymous. If this is not the case, I would like to hear about it.

Intervention:
Here I foresee a very complicated technical discussion as the concepts of predictability, experience and familiarity are in the same atmosphere.

Intervention:
I would like to join the enthusiasm of professor Schell about the application of biotechnology. I would also like to stress a question asked by somebody else here: What will happen if we do not use biotechnology? I would say we are at risk losing genetic diversity which we are also trying to preserve by other means.

SETTING THE SCENE FOR ECOLOGY

LONG TERM ECOLOGICAL IMPACTS OF THE RELEASE OF GENETICALLY MODIFIED ORGANISMS

Dr. M. J. Crawley, Department of Biology, Imperial College, United Kingdom

INTRODUCTION

The emphasis of this paper is on the kind of ecological process that might lead to unintentional, negative long-term impacts on ecosystems resulting from the introduction of GMOs. For this purpose, I have adopted a simple, pragmatic definition of 'long-term': a process is long-term if its characteristic time scale is of the order of 10 to 100 generations. A long-term ecological study, however, should last for a minimum of 10 years.

You might ask "What's all the fuss about?" After all, humans have been introducing novel genotypes into ecosystems since time immemorial. Some of the issues are listed in Table 1. If any one of these fears were to be realized, it would be reason enough to rue the introduction of the GMO that caused the problem. Taken in aggregate, they present a catalogue of horrors which, given the level of our ignorance about the precise consequences of introductions might appear to give the opponents of the introduction of GMOs an unanswerable case.

Table 1.

WHAT'S THE WORRY? SOME OF THE ISSUES THAT HAVE BEEN RAISED AS REASONS WHY GMOs SHOULD NOT BE INTRODUCED INTO THE ENVIRONMENT
build-up of resistance to valuable pesticides
creation of new agricultural weeds
creation of new crop pests
replacement of K-strategists by r-strategists
invasions of natural habitats
loss of biodiversity
loss of genetic diversity within valued species
physical obstruction of human access
increased rates of soil erosion
interference with nutrient cycles

My purpose here is to look at what we *do* know about the ecology of long-term processes, in order to assess the probable consequences of introducing GMOs of different kinds. You will not be surprised to learn that my answer is "It depends". All of the introductions of GMOs to date have passed off without incident in the short term and are almost certainly not going to have adverse long-term environmental consequences. But equally clearly, we can design (even if we can't yet make) GMOs that would have the most catastrophic long-term impact on ecosystem function. This, surely, is the nub of the dilemma. If we allow ourselves to be satisfied with risk assessment work that was carried out with organisms that could never be anything other than safe, then we are asking for trouble later on, when combinations of species and construct are developed that pose environmental threats of an entirely different kind.

Perhaps the main cause for concern about long-term effects is the insidiousness of that most ecological of all processes: exponential growth. Given that there is a reasonably high threshold level at which we would regard a population as having become 'problematical', then the problem is likely to be recognized so long after the original introduction that any residual concern associated with the release itself will long since have evaporated. Likewise, the population will have built up such a momentum, such an ecological head of steam, that by the time the problem is perceived, and resources are mustered to counter it, the problem will have gone completely out of hand. In the figure below, the population does not break through the 'nothing's happening' threshold until 100 years after the release.

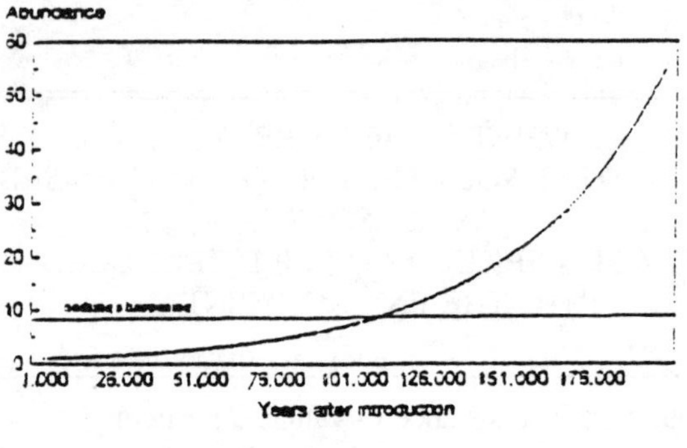

ECOLOGY

There is no widespread public understanding of what ecology is about. To most, I suspect, it is a vague notion of something to do with Green Politics, with recycling aluminium cans, or with a philosophy of 'sustainable growth' (an oxymoron, surely?). To ecologists, however, ecology is the scientific study of the distribution and abundance of organisms. Prosaic, but precise. Good introductions can be found in Colinvaux (1973) and Begon, Harper and Townsend (1990), while more specialised aspects are dealt with by Elton (1958), Pielou (1969), MacArthur (1972), Maynard Smith (1974, 1982) and Crawley (1986, 1992).

The object is to determine the factors that determine why species are found where they are, and not where they aren't. It is to understand what determines the characteristic level of abundance and the degree of temporal variation exhibited by different species. In addressing these issues, the ecologist is concerned with 5 big themes: (1) a place in which to live; (2) food; (3) competitors; (4) mutualists and (5) natural enemies (Andrewartha & Birch 1954, Bender, Case & Gilpin 1984, Hairston 1989). Most of the ecological risk assessment work on GMOs that has been carried out to date has gone no further than this. For example, since maize can not survive the British winter, it fails on question (1): it does not have 'a place in which to live', and so the introduction is safe. Slightly more advanced thinking is involved with the issue of introducing transgenic oilseed rape. Since it is a crop plant it is not competitive with native vegetation, and hence it fails on question (3): its competitors are so effective that there can't be a problem.

The truth of course, is somewhat more complicated than this. All species have the ability to increase under *some* kinds of conditions (reproduction is one of the defining features of living organisms), so if the conditions occur in certain places, then there may be a refugium from apparently hostile conditions, and the organism does, after all, have 'a place to live'. For example, an insect species which is quite incapable of overwintering under field conditions, might maintain residual populations by overwintering in heated buildings, from which field habitats are recolonized each spring.

Again, one of the most widespread misconceptions about ecology is that 'competitive ability' is a species-specific trait (like territoriality, longevity, or maximum number of eggs). Competitive ability depends crucially on the nature of the location (pristine or disturbed, early or late successional), the kind of year it is (wet or dry, hot or cold), the identity of the competitors, their condition (e.g. their growth rates) and the condition of the organisms in question (e.g. their parasite burdens). Competitive ability is highly context-specific, and it could be a grave mistake to assume that an introduced genotype would be outcompeted in sufficient of the habitats into which it might find its way, that its persistence or increase was impossible.

Another major difficulty when considering the long term ecological implications of GMOs is the question of spatial scale (Table 2). We need to consider processes that occur extremely quickly and at extremely small scales (like nutrient uptake by mycorrhizal fungi) and other processes that occur extremely slowly at very large spatial scales (like sedimentation in growing salt marshes).

Table 2. Spatial Scale

SMALL SCALE	LARGE SCALE
interactions between micro-organisms and their neighbours	erosion and sedimentation
nutrient uptake by plant mycorrhizae	migration of birds and mammals
infection of insects by pathogens	nutrient cycles within watersheds

The only recourse in the face of this kind of complexity is to make simplifying assumptions. The assumptions most often made by ecologists are these: pick the spatial scale on the basis of the problem in hand; assume that state variables which are influenced by relatively slow processes are constant for the purposes of study; assume that relatively fast processes are in equilibrium. This allows us to apply the principal of parsimony, and to arrive at a tractable framework for tackling the problem. On the other hand, these are assumptions and so they might be wrong. A fast process might not be in equilibrium, and something we thought would be constant might actually turn out to be changing quite rapidly. Ecologists are trained to seek simplicity, then distrust it. They are also taught to expect the unexpected. That's what makes thinking about the consequences of GMOs so challenging and so difficult. At one level there is absolutely no cause for concern. At another level, there is the knowledge that it could all go horribly wrong.

GENOTYPE AND PHENOTYPE

One of the major difficulties of this whole debate is that there is no way of predicting the ecological behaviour of the phenotype from a knowledge of the genotype. Phenotype is so dependent upon complexities that occur during development (i.e. before birth) and so affected by environment (i.e. by history and experience) that it is unlikely that we shall ever be able to predict the precise consequences of small genetic changes on population-level behaviour. It is one thing to be able to predict that plants with luciferase genes will have a phenotype that glows in the dark. It is quite another thing to predict the degree to which the expression of an insect tolerant phenotype resulting from the insertion of Bt genes will affect the net reproductive rate of a plant in a given habitat.

SINGLE-SPECIES DYNAMICS

The only way to approach these fundamental issues is to begin with the basics. The most basic ecological principle of all is called the **invasion criterion**. This says that for every extant species, there is a combination of ecological conditions (see below) under which the species exhibits the **ability to increase when rare**. In the jargon of population dynamics,

$$\frac{\delta N}{\delta \tau} > 0 \textit{ for } N \textit{ small}$$

Were it not for this criterion, then species would go extinct following a series of catastrophes; the ability to bounce back after catastrophic reduction in numbers is vital for persistence. The invasion criterion is also widely used in evolutionary ecology to ask questions about the fitness of individuals possessing certain traits. A mutant trait can increase to fixation only if the phenotypes carrying that mutant can increase when rare; this idea is at the heart of the Evolutionary Stable Strategy, which states that a given trait is an ESS when that trait can not be invaded by different genotypes (i.e. the non-ESS genotypes fail the invasion criterion; Maynard Smith 1982).

The obvious corollary of the invasion criterion is that populations will tend to grow exponentially (at least when population densities are relatively low):

$$\frac{\delta N}{\delta t} = rN > 0 \rightarrow \textit{exponentialgrowth}$$

but, inevitably, limits to growth occur, and hence

$$\frac{\delta N}{\delta t} \rightarrow 0 \textit{ as } N \rightarrow K$$

where K is equilibrium population density (sometimes rather confusingly called the 'carrying capacity', since densities can exceed K, and when they do, population density tends subsequently to decline; May 1973).

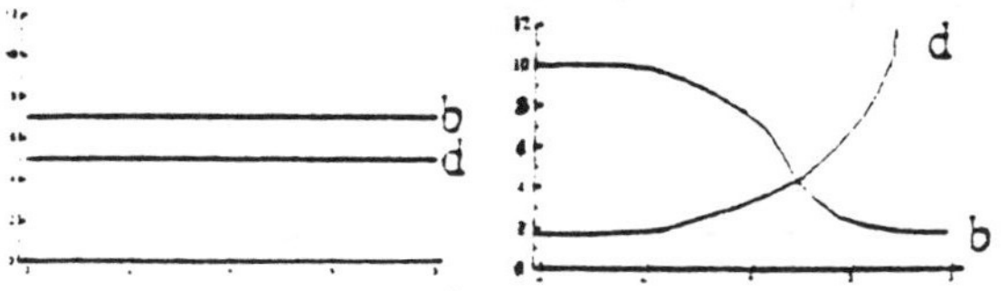

Density independence Density dependence

Now the ecological interest is focused on these 2 parameters: r (the intrinsic rate of increase) and K (the equilibrium density), and we can begin to ask questions about how genetic engineering might affect the mean values of these parameters and their variances. Variability in both r and K can have profound effects on population dynamics, because between them they affect the likelihood of population overshoot at high densities (and hence the prospect of cyclic or chaotic dynamics; see below), and perhaps more importantly, they determine the probability of local extinction (e.g. when low K coincides with temporarily negative r).

The existence of an equilibrium population density, K, depends absolutely upon the action of one or more **density dependent factors:** these are processes which cause the rates of birth (b), death (d) or dispersal to change as population size increases or decreases. Broadly speaking, a density independent population would either decline inexorably to extinction (if d > b), or rise exponentially (if b>d). If there is density dependence, then there may be an equilibrium; it will occur at that density where the gains from birth and immigration are balanced by the losses from death and emigration. In the real world, of course, the rates of birth, death and dispersal vary from year to year and from place to place within a year, and so we should not expect to find any crisp determinism in the magnitude of the equilibrium population density, or any clear pattern in the level of variation about that equilibrium.

While the existence of the equilibrium depends absolutely on the operation of density dependent processes, the size of the equilibrium is determined by the interaction between all the processes (density independent as well as density dependent) that affect the vital rates (birth, death and dispersal). Suppose that the birth rate declines as population density increases because of a shortage of food for each reproductive female (intraspecific resource competition). Now if the death rate is high and density independent (dotted line) we have an equilibrium density N_1, and if the death rate were to fall (solid line), then equilibrium density would rise to N_2. Without a precise knowledge of the nature of the density dependence (negative in this case) and the magnitude of the change in density independent factors, it is impossible to predict the change in equilibrium population size. Moreover, even *with* this knowledge, it would not be possible to predict the transitional (i.e. medium term) dynamics by which the population changes from its present state to the new state. For example, random events might mean that the population goes extinct during one of the lows of transitional fluctuations, even though the parameter values are those of a high, stable equilibrium population.

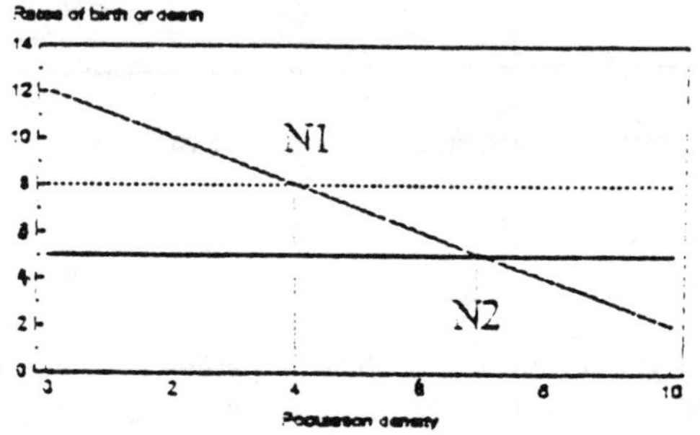

In this simple scheme of things, the regulating factors were assumed to work continuously and in a roughly linear manner. But this is not typical of the behaviour in real ecological systems, where thresholds and non-linearities are commonplace. Both these phenomena present considerable difficulties for assessing the long term risks of introducing GMOs, largely because of the uncertainties they introduce. For example, does the present lack of any adverse response mean that the introductions that have been carried out so far are inherently non-problematical, or might there be cumulative effects but we have not noticed anything because the threshold has not yet been passed. Similarly, if an ecosystem response is strongly non-linear, then small changes which made no difference during the risk assessment phase might make enormous differences later on when the introductions were made on a much larger scale. A good example of threshold ecological processes are the cumulative effects of acid rain on soil chemistry. These lead to strongly non-linear behaviour because the buffering capacity of the soil changes in a threshold manner, so that the solubility of toxic heavy metals (like aluminium) changes abruptly with declining soil pH with sudden, and severely debilitating effects on plant growth. This kind of ecological behaviour is worrisome, because it suggests that environmental impacts estimated over one range of conditions may be of little use in predicting behaviour over a different range of conditions (i.e. that small-scale ecological risk assessment experiments may be fundamentally flawed).

PATTERN OF DYNAMICS

Does complex ecological behaviour require complicated models to represent it? The answer is a resounding No! The pattern of dynamics exhibited by a population in a constant environment depends upon the nature of the density dependence:

$$N_{t+1} = N_t \text{ x net reproductive rate x } f\,(N_t, N_{t-1}\, N_{t-2}, \ldots.)$$

where the population at various times in the past influences current reproduction via the effects of body condition, size, parasite burden, sex ratio, etc. (as embodied in the function f). The simplest possible form of this model has linear density dependence and only a single time-lag:

$$N_{t+1} = \lambda N_t(1 - N_t)$$

Despite its simplicity, this 1-parameter model is capable of exhibiting the full range of dynamic behaviour from stable point equilibrium, through damped oscillations and stable limit cycles, all the way to chaos (May 1973). What this means, in the present context, is that small changes in parameter values caused by genetic engineering, could have dramatic effects on population dynamics. Without an understanding of the shape of the recruitment curve, the consequences of a given change in phenotype could not be predicted. Even in the unlikely event that we did know the consequences of genetic engineering on the parameter values, the outcome would still be completely unpredictable if the dynamics were to lie in the chaotic region, because of the fundamental property of chaos: **extreme sensitivity to initial conditions.**

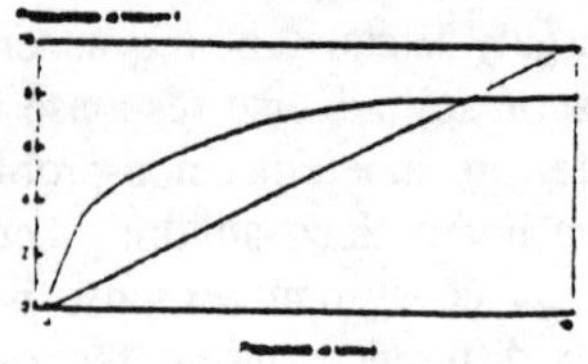

A recruitment curve plots the size of the population in the next generation as a function of the size of the population this generation. Constant populations lie somewhere along the replacement line (a straight line with unit slope). The precise location of any equilibria depends upon the shape and location of the recruitment curve; populations increase when the recruitment curve lies above the replacement line, and decline when they lie below it. An intersection between the two lines is an equilibrium population density. The equilibrium is unstable if the recruitment curve cuts the replacement curve from below, and stable if it cuts it from above. The stability of the equilibrium depends upon the kind and intensity of density dependence, as reflected in the precise shape of the recruitment curve at the point of intersection. If this shape is known, then in principle, the dynamics can be predicted. If the slope of the recruitment curve is shallow (the case below), then the population will show a rapid return (either exponentially or with damped oscillations) to a stable point equilibrium.

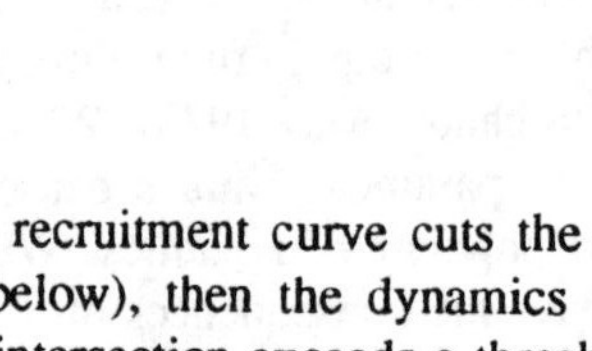

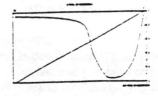

If, on the other hand, the recruitment curve cuts the replacement line with a steep, negative slope (the case below), then the dynamics will be complex (they will be chaotic if the slope at the intersection exceeds a threshold value).

Finally, the recruitment curve may suggest the existence of multiple equilibria. In the graph below the left most equilibrium is unstable; a GMO which caused population density to fall below this level would lead to the extinction of the population.

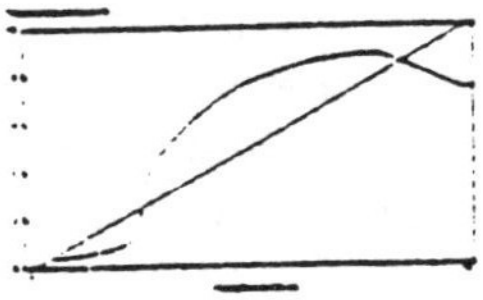

The important point is that small changes in parameter values can have dramatic effects on population behaviour. In the figure below, changing the value of λ from 3.0 to 4.0 causes the dynamics to change from a stable, low-amplitude, two-point cycle into chaos: random-looking dynamics that are completely deterministic, but which show extreme sensitivity to initial conditions. It is completely impossible to predict population size 10 or 20 time periods ahead, even though there is no environmental 'noise' in the system. Real ecological predictions are doubly difficult, because spatial and temporal environmental stochasticity would be overlain on any inherent chaotic tendencies that the system might possess.

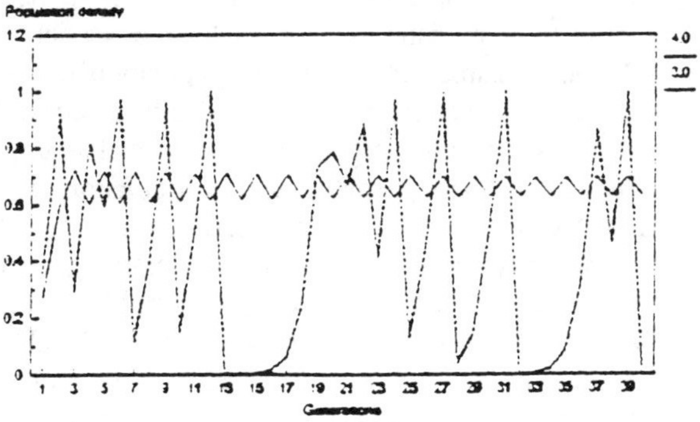

SPATIAL DYNAMICS

In considering long-term temporal dynamics we are drawn inevitably into the study of spatial dynamics; the study of range expansions and contractions, and the details of the dynamics of local patch-occupancy. The topics involved can be gathered under 3 headings:

1) Spread from a point source (diffusion z & population growth r) with an 'invasion front'.
2) Point invasions from numerous foci (pattern & size).
3) Metapopulations (dispersal & extinction).

Each of these is a study area in its own right, but data from all 3 have been used to contribute to the debate about the introduction of GMOs, mainly in so far as they relate to the Introduced Species Model of biological invasions (ISM, see Elton 1958 and below). At this point, all that we need to note is that at least one new parameter needs to be added to our simple model for population dynamics: this parameter (z) is analogous to the diffusion coefficient of physics, and along with the intrinsic rate of increase (r) the two parameters determine the annual rate of advance of the invasion front across formerly unoccupied territory. The rate of radial advance (km.yr) is more or less constant for many models of invasion (proportional to the square root of the product of the intrinsic rate of increase and the diffusion coefficient) and for most of the real field examples for which good data can be gathered (see Skellam 1951). Models of multiple point invasions are less well developed, but well studied field systems do exist (e.g. the alien grass *Bromus tectorum* in the north west USA; Mack 1981).

The dynamics of metapopulations of GMOs are modelled most simply by considering the world as consisting of a patchwork of more or less isolated units of habitat which, at any given time, can be either occupied or vacant. The dynamics then consist of the series of states, summed over all patches, as occupied patches become extinct (the food runs out, natural enemies become abundant, or a density independent catastrophe wipes out the local population) and as formerly unoccupied patches are colonized by dispersing individuals that have emigrated from occupied patches. All this is eminently realistic (if you make the spatial scale sufficiently small, then all population dynamics must be metapopulation dynamics; for sessile organisms like rooted plants, for example, each death is equivalent to a local extinction). Some ecologists, however, argue that if dispersal is important, then you are studying the system at the wrong spatial scale. Populations should be studied, they would say, at a spatial scale which is sufficiently large that dispersal is irrelevant. If you do not study the population at this large scale then you can not hope to discover the identity or the mode of operation of the regulating factors, because these will be invisible at the smaller scale on which local extinctions occur frequently. This all matters in the present context, because the introduction of GMOs is likely to be frequent, irregular and spatially heterogeneous. Even though it might be the fate of most introductions to fail, the GMO could persist by dint of efficient dispersal coupled with a high rate of reproduction in that small subset of patches where conditions were favourable.

TWO-SPECIES INTERACTIONS

Two species interactions are an abstraction of the same kind as single-species systems, but they allow us to concentrate on the most important elements of ecological behaviour. It is natural to classify 2-species interactions into 5 kinds based on the trophic levels of the actors involved: (1) plant-resource; (2) plant-herbivore; (3) predator-prey; (4) disease-host; and (5) host-mutualist. The first 4 have in common that they are essentially 'exploiter-victim systems' and we might expect their population dynamics and evolutionary biology to have certain broad features in common. The last category, mutualist-host, is a rather special case, since it is benefit rather than cost that one species contributes to the other. Theoretical models of mutualism tend to turn, in Bob May's immortal phrase, into 'an orgy of mutual benefaction'. In fact, most of the benefits of mutualists are of the diminishing returns kind, and the benefits tend often to be asymmetrically distributed between the interacting species. It may therefore be preferable to model the absence of mutualists as being potentially harmful, rather than always viewing the presence of mutualists as beneficial.

The theory of exploiter-victim interactions is well developed (Crawley 1992) but not to the point at which predictions could be made confidently about the consequences of genetic engineering. Suppose, for example, that it was proposed to introduce a genetically engineered insect into the country as a control agent for a pernicious alien weed. The non-transgenic insect is a pest of related plants in Europe which has been engineered to exhibit strict monophagy for the target weed. In order to predict what would happen we should need to know the following list of attributes:

(a) the intrinsic rate of increase of the weed;
(b) the carrying capacity for the plant in the absence of herbivores;
(c) the functional response of the herbivore (the way that food intake per insect depends upon the amount of food available per unit area);
(d) the intrinsic rate of increase of the herbivore;
(e) the per-capita impact of herbivore feeding on plant performance;
(f) the nature of density dependence in the herbivores' numerical response (the way that herbivore reproduction depends upon food availability and food quality);
(g) the extent of the overlap in the spatial habitat use of the plant and the insect;
(h) the relative dispersal rates of the two species.

It will be immediately clear that there is no prospect of being able to predict the effects of genetic engineering on more than one or two of these parameters. And this supposes that the traits are constants whereas in reality they will vary from season to season and from place to place.

This should not be seen as a capitulation to the Many, Complex and Interacting school of ecology. On the contrary. It is a question of being clear about what it is reasonable to expect an ecologist to be able to predict. Given that ecologists don't profess the ability to predict next year's weather, it is disingenuous to ask them to predict next years population dynamics (which depend on next year's weather and on much else, besides). There are, however, a great many generalizations which are sufficiently robust that they can alert us to the most problematical classes of GMO.

Take the example of herbivores. What kind of genetically engineered herbivores are likely to have greater and lesser ecosystem impacts (Table 3).

Other things being equal, it would be safer to introduce a transgenic insect than a transgenic mammal. This assertion is borne out by the Introduced Species Model;

compare the devastating impacts of introduced mammalian herbivores (rabbits, goat, muskrat) with the substantially less serious consequences of introduced insects (Colorado beetle, knopper gall insect, Japanese beetle; Elton 1958). A corollary of this prediction relates to the species richness of the target ecosystem. The proportional impact of small, sedentary monophagous herbivores is likely to increase as species diversity declines. In the limit, a monoculture of the food plant of an insect herbivore is more likely to be affected by the introduction of that insect than by the introduction of a more polyphagous mammal (but the impact on neighbouring habitats would need to be carefully considered, since here the strength of the effects would be reversed). Similar arguments can be made about the introduction of transgenic natural enemies (e.g. for the control of pest insects; see Table 4).

Table 3. Essence of plant-herbivore interactions

BIG IMPACT	SMALL IMPACT
Large Mobile Polyphagous e.g. ungulates	Small Sedentary Monophagous e.g. insect herbivores

Table 4. Degree of impact of natural enemies

BIG IMPACT	LESSER IMPACT
High virulence Efficient transmission Rapid numerical response Aggregation at high prey density Linear functional response	Low virulence Demanding transmission (e.g. vectors) Numbers limited by other factors Inflexible spatial distribution Saturating functional response

TWO-SPECIES INTERACTIONS

As with single-species systems, the dynamics of coupled populations depend upon the nature of the density dependence. In a plant-herbivore system, for example, the herbivore might increase when plants were abundant, and the plant population might decline when herbivores were numerous, but density dependence in the plant population (e.g. competition for light at high biomass) and in the herbivores (e.g. limited territory size, irrespective of plant biomass) might imply the existence of a stable point equilibrium (a). The contours separating the regions where the plant increases or decreases and the herbivore increases or decreases are called the 'zero-growth isoclines' for the two species. Depending upon the shape of the isoclines, the point at which they intersect could be a stable point equilibrium (like point (a), or it could be the focus of a stable limit cycle. In the case of a stable point cycle like (a), small departures in plant or herbivore density would be damped out with a rapid return to equilibrium. Despite the existence of the equilibrium, however, this model system is not globally stable. There are still circumstances (like b) where immigration of herbivores can drive the plant population to extinction, and others (c) where plant abundance falls so low that the herbivores go extinct despite the fact that the plant is increasing in abundance. Thus, proof of the existence of an equilibrium does not consititute proof of the persistence of a system. Global stability (resilience independent of the vicissitudes of the environment or the ups and downs of internal population dynamics) requires the existence of refugia for both populations (the plant and the herbivore in this example). Generally speaking, these refugia can only exist in the form of immigration from outside the boundary of the local system and in this sense, global stability is virtually impossible in closed systems.

Now it is a truism that complex systems like ecosystems tend to be open rather than closed (see below). But a lot of human-influenced landscapes are becoming closed to dispersal, as patches of habitat are destroyed which might otherwise have formed the source of propagules by which this rescue effect could work. Such fragmented systems which are closed to dispersal are very unlikely to be globally stable.

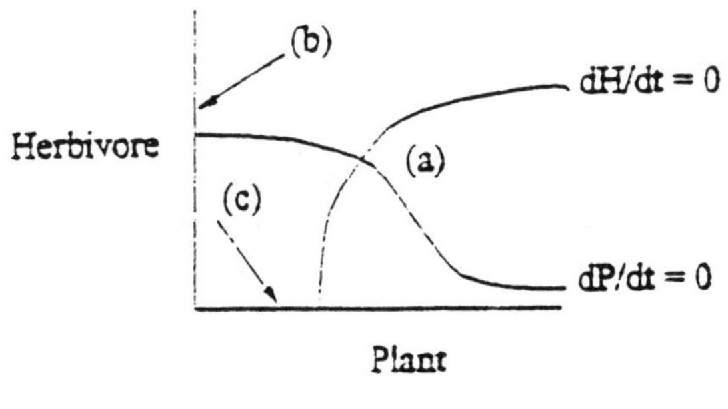

MULTI-SPECIES COMMUNITIES

There are 4 elements that we need to consider in looking at the community-level effects of introducing GMOs:

1) Number of trophic levels.
2) Number of species per trophic level.
3) Number of links within and between trophic levels.
4) Strength and directions of interactions.

Community-level function will be influenced by the dynamics reflected in points (3) and (4) and these, in turn, will be affected by GMOs to the extent that their introduction leads to an alteration in the number, strength and direction of linkages. Perhaps the most important consequences are those involving the relative strengths of positive and negative **feedback loops.**

The location of the GMO in relation to the influential feedback loops is vital. Any GMO which amplified positive feedback loops or weakened negative feedback loops could have disproportionate long-term impact on the ecosystem. We need to know what the feedback loops are, and what are the direct and indirect effects of the GMOs on them.

THE NATURE OF ECOSYSTEMS

It is important, therefore, that we know whether ecosystems are:

1) Closely coupled, highly integrated organisms? (e.g. Clemments 1924), or;
2) Loosely coupled, randomly assembled subsets from a pool of adapted species? (e.g. Gleason 1926).

This question matters, because if ecosystems are of the first type, then effects resulting from the introduction of GMOs will be profound and far-reaching. If ecosystems tend to be of the second type, then the effects of GMOs are likely to be local and ephemeral. The nature of ecosystems might be summed up in these two contrasting quotations, one from a Victorian poet and the other from a pioneering American plant ecologist:

Francis Thompson (1859-1907), The Mistress of Vision

All things by immortal power, near or far, hiddenly
To each other linked are,
That thou canst not stir a flower
Without troubling of a star

H.A. Gleason (1926), The Individualistic Concept of Plant Association

The community is not an organism
but merely a coincidence

The modern consensus among ecologists is that Gleason was closer to the truth, and that many of the linkages in ecosystems are weak and asymmetric. The issue of whether much of the biodiversity in ecosystems is functional or redundant is still a matter of intense debate. Some experiments show that more diverse systems recover more quickly and more completely following severe stress (e.g. grasslands recovering from drought). On the other hand, long-term, blanket elimination of soil organisms using pesticides has not had catastrophic effects on plant productivity, as some environmentalists had predicted. The theory available on this point is highly equivocal. Some models say that increasing diversity increases stability (the argument being 'don't put all your eggs in one basket'). In contrast, other models say that increasingly complex systems are less stable (the argument being 'if something can go wrong, it will'). We just don't know what most species do in most ecosystems. The only way to find out is to do controlled, replicated, long-term removal experiments. Likewise, we really can't predict what will happen when we release a new GMO into an ecosystem. Here again, the only way to find out is to do controlled, replicated, long-term addition experiments. There simply is no alternative. If we want to know, then at least we know how to find out. However, it is not at all clear to me that there is any genuine wish to know the long term ecological consequences of introducing GMOs. The attitude still prevails among decision-makers that 'we have thought really hard about this, and we can't see any problem'. The ecologist says something only slightly different: 'a problem is extremely unlikely to occur, but if does, then you'll regret it in a big way'. As US Office of Technology Assessment said, there are "reasons to continue to be cautious, but there is no cause for alarm". What has not been generally appreciated, is that in order to be genuinely cautious (rather than simply to play lip service to pre-release assessment) will require long-term investment in studying the environmental impact of a wide range of transgenic constructs in a variety of habitats, world wide.

KEYSTONE SPECIES

Some species are more important than others. Unfortunately, it is not always obvious which species these are. Few casual observers visiting the Chihuahuan deserts of the south-western USA, for example, would predict that removal of the tiny nocturnal rodents known as kangaroo rats would cause the system to change completely (long term studies by Jim Brown and his colleagues have shown how the system changed from open desert into scrubland following removal of the rodents; Brown 1986).

The definition of keystone species is pragmatic: a native species is keystone if its experimental removal causes major ecosystem-level changes in structure, dynamics or nutrient flows. Likewise, introduced species are keystone if their introduction causes major, ecosystem-level change. Introduced animals have often taken on the mantle of keystone species. Rabbits were introduced into the British Isles as domestic livestock for human consumption but have escaped to become the dominant herbivores in most lowland ecosystems, capable of preventing succession to oak woodland. The introduced tree Myrica faya has become a keystone species on the lava lows on the main island of Hawaii (Vitousek & Walker 1989).

Clearly, GMOs will have greater impacts when their introduction affects keystone species than when the target species if there is one) plays a lesser ecosystem role. The lesson we learn from Jim Brown's work, however, is that the keystone species in most ecosystems are unlikely to be known, because the experimental removals will not have been carried out, and you can not deduce the identity of a keystone species from theoretical considerations. In some cases we are forewarned, however. For example, the proposed northwards extension of the range of sea bass, a top marine predator, by introducing an antifreeze gene is patently outrageous. A top predator like the sea bass is likely to be a highly influential species, whose introduction to a new habitat would have far-reaching consequences at several lower trophic levels. Such an introduction simply should not be contemplated. We can't be sure that sea bass is a keystone species, but there is more than reasonable doubt.

MULTIPLE STABLE STATES

Systems with strong linkage, high connectance and widespread asymmetry of interactions hold the potential to exhibit multiple stable states. Systems containing keystone species for example, tend to have at least two more-or-less stable states: one with the keystone species, and a very different one without it. Generally speaking, the things we know about multiple stable states come from resource systems like fisheries that have been overexploited by humans. The problem is that the switch between different stable states may be irreversible (or at least extremely difficult to reverse except in the very long term) and Crawley's 'rule of multiple stable states' says that the stable state into which a system flips is always less desirable than the state it flipped out of. These large, long-term impacts of GMOs at the ecosystem level are only as likely as the effects of GMOs on keystone species. Such impacts are unlikely, but they may prove to be extremely difficult to anticipate because of the difficulty of identifying a keystone species before it disappears (or before an alien keystone species becomes established). Recall that the only way we know the identity of keystone species is by long-term field experimentation (addition or removal); there is no theoretical means by which a keystone species can be identified from first principles. This is because species are (or are not) keystone because of **parameter values** rather than as a result of qualitative, structural aspects (traits) of the ecosystem.

LESSONS FROM INTRODUCED SPECIES

The Introduced Species Model has suffered from a degree of abuse in the debate about introducing GMOs. It was absurd to equate the risks of introducing transgenic lines of familiar crop plants with the risks of introducing brand new alien species whose ecology in their new environment was unknown and almost totally unpredictable. Equally, it was wrong to say that nothing at all about the likely outcome of invasions of alien species could be predicted in advance.

Alien plants in the UK

The lessons learned from alien plants in the UK can be grouped under 3 sub-headings.

(1) We do know that some habitats are more invasible than others. The invasible habitats are frequently disturbed, have low percentage cover of perennial vegetation, and are close to human habitation. The less invasible habitats tend to be remote from human influence and to be more pristine, less disturbed and to have a high percentage cover of perennial vegetation.

(2) We know that species belonging to some taxa of plants are much more likely to be invasive than others (invasive families include Cruciferae, Labiatae and Caryophyllaceae, while non-invasive families are typified by Polypodiaceae, Orchidaceae and Cyperaceae; note, however, that Sod's Law places the world's worst weed, *Cyperus rotundus* in this latter family).

(3) The so-called 10-10 rule (Williamson 1993) is wrong for plants. Far fewer than 1 in 10 of the plants brought to Britain has become established and of those that become established, less than 1 in 10 become pests (there are roughly 11 alien species that are sufficiently abundant to be classified as weeds in the UK out of more than 25,000 plant introductions; a rate of less than 1/20 of 1%).

How good is the Introduced Species Model?

The introduced species model is as good as our pre-release experience of the organism in question. In cases where we have long experience of the species, the outcome of the introduction is likely to be predictable. Similarly, confidence in our prediction of the outcome is increased as the number, size and frequency of introductions is increased (e.g. a species predicted to become widespread and abundant may fail to establish from a few small-scale introductions just through bad luck). Finally, the degree and quality of pre-release screening will influence the quality of the predictions; a well-screened release is much less likely to become problematical than is a random introduction.

Can the impact of introductions be predicted?

Yes and no. We can be almost certain that the introduction of a pest species will lead to that species becoming a pest in the country of introduction (this is one of several biogeographic manifestations of Sod's Law). Likewise, we can be highly confident that a well-screened insect biocontrol agent introduced to control a weed will **not** become a nuisance for native flora and fauna (the safety record of weed biocontrol using insects is excellent; out of more than 1000 introductions, there is not a single report of a problem (e.g. the introduced agent switching to attack native or valuable plant species); compare this with the appalling record of casual introduction of vertebrates for pest control, like cane toads in Australia).

Conversely, we have very little ability to predict the consequences of introducing little known or unscreened, non-pest organisms. They might take off (e.g. knopper gall insect; little owl, collared dove) or they might not (eagle owl, most garden plants, swallow-tail butterflies). The point is that there are no traits and no suites of parameter values that can be used reliably to separate the successful from the unsuccesful invaders in any usefully predictive sense. There are broad patterns (e.g. species that are widespread and abundant in the native range, and species with high potential rates of increase are more likely to establish) but they often have negligible predictive power for practical purposes.

DIVERSITY

The most important ecological concept involved in considerations of diversity is the **competitive exclusion principle:** this states that under any given, constant set of conditions, one species will be superior and, given long enough, will displace all others. The expectation, therefore, is of uniformly low species richness (i.e. monocultures) within each type of habitat. While this does occur in natural habitats (e.g. bracken stands, reed beds, heather moorland), the norm is for the coexistence of several to many species in a given guild (e.g. grasslands typically support 3 to 45 species of vascular plants in a 10m x 10m area; desert rodent guilds have 2 to 6 species differing in body size and diurnal behaviour patterns).

So the question now becomes: Why does competitive exclusion not happen? There are a range of possibilities (Chesson 1986; Roughgarden 1986) including:

1) limited niche overlap (i.e. competition is not as severe as it might first appear);
2) prevention of dominance (e.g. by selective attack by natural enemies);
3) random processes (conditions are not constant for long enough for dominance to occur);
4) non-congruence of spatial distributions (refugia for the inferior competitors);
5) different phenology (the species are not active at the same times of year);
6) environmental variation (conditions favour different species in different years).

Genetic engineering could affect any of these processes. For example, niche breadth (and hence niche overlap with potentially competing species) might be altered by genes for diet selectivity (e.g. transgenic biocontrol agents). The GMOs will reduce biodiversity to the extent that they enhance the performance of the competitive dominant, or eliminate other species directly or indirectly (e.g. a GMO might eliminate an out-breeding plant species by attacking its specialist pollinator).

To put these kinds of problems into perspective, we might develop the last example in more detail. Of course it is possible that the demise of a specialist pollinator could be a problem for an outbreeding plant species, but it is extremely unlikely to be a problem for the following reasons:

a) most outbreeding plants do not rely on one specialist pollinator;
b) most plants (even obligate out-breeders) are not pollinator-limited;
c) plants which are pollinator-limited tend to have one or more forms of insurance against pollination failure (e.g. a large bank of dormant seeds in the soil);
d) there are no documented cases of this kind of thing happening.

This is not an argument for complacency, but simply a plea that we use the information that we do possess to weigh the uncertainties involved in assessing the likely long-term consequences of particular transgenic introductions.

BIODIVERSITY

One of the most frequently expressed fears about the introduction of GMOs is that this will lead to reduced biodiversity. There are a number of important ecological issues here. First of all, we need to know whether species richness is maintained as an equilibrium between extinction and establishment (like large-scale metapopulation dynamics), or is maintained by the non-equilibrium failure of competitive exclusion. For instance, it is easy to build theoretical models that allow many species to coexist so long as the environment is sufficiently variable in time and heterogeneous in space that the inferior competitors are always able to find some refuge, created by chance alone, in which they are freed from domination by the superior competitors.

Descriptive theories of biodiversity tend to rely on physical disturbance or changes in productivity to alter the rules of competitive exclusion. The intermediate disturbance hypothesis (Grime 1979, Connell 1979) says that with low disturbance the dominant can achieve competitive exclusion and diversity will be low. With high disturbance, the size of the pool of species capable of living under such a severe regime of disturbance is small, and so diversity, once again, will be low. Only in circumstances where disturbance is of intermediate intensity will the pool of species be large enough, and the likelihood of dominance low enough, that species diversity will be high. It is easy to parody this argument, however, and to erect the 'intermediate anything hypothesis'; too much or too little of any ecological factor will mean that only a few species can tolerate the extreme conditions at either end of the x-axis.

The productivity hypothesis (Oksanen et al. 1981) has to do with trophic-level biodiversity, and says that in low productivity systems (like deserts, plant productivity is so low that herbivores are scarce because the resources they need are simply too dilute to exploit). In moderately productivity environments, plant productivity is sufficient to support herbivores, but herbivore productivity is insufficient to support natural enemies. This means that herbivores are abundant (they are food-limited) but predator numbers are low. In high productivity systems, the productivity of herbivores is sufficient to support predators, so paradoxically, herbivore abundance is lower than in systems of intermediate productivity. This is because in high productivity systems the herbivores are enemy-limited rather than food-limited (enemy-limited populations are bound to be held at lower population densities than food-limited populations). These views are controversial, and it is not clear whether the introduction of GMOs would alter the characteristic disturbance rate of an ecosystem or its level of primary productivity, to the extent that indirect effects on biodiversity through either of these routes is likely.

At present, the impact of GMOs on biodiversity tends to be seen from a much simpler, short-term perspective: biodiversity will be reduced by GMOs which are themselves the competitive dominants in their guilds (i.e. the GMO is itself a keystone competitor and the direct cause of competitive exclusion), or by GMOs which are catastrophic agents of mortality for their host organisms, and their hosts, in turn, are keystone species (i.e. the GMO is the direct cause of local extinction through increased mortality or reduced fecundity of its host (or prey) species).

SUCCESSION

GMOs might change several of the aspects of community succession; they might alter its direction, change its rate, or influence its endpoint. Effects on the course of succession are more likely if the GMO has impacts on the keystone species in one or more of the stages of succession than if they influence non-keystone species in only one stage. There are 3 models of the way that succession works (Connell & Slatyer 1977):

1) Facilitation: species A changes conditions so that species B can invade.
2) Inhibition: species A prevents the invasion of species B.
3) Life-history: species B replaces species A, simply because B is longer-lived.

The modern view, however, is that each species in succession tends to have its own rules for invasion and eventual replacement (Bazzaz 1979; Boucher, James & Kesler 1984; Grubb 1986). Some interactions will involve facilitation, but others in the same succession will involve inhibition or life-history. The role of GMOs in altering successions has not been considered in any detail, but introduced species can have the most profound effects in changing the course of succession and in altering the identity of the eventual dominant species. For example, the introduced nitrogen-fixing tree *Myrica faya* short-circuits entire primary successions on the island of Hawaii, with dire consequences for the biodiversity of native species that would normally occur in the multi-step succession from bare larva to mature rain forest (Vitousek & Walker 1989). One way in which a GMO might have a major impact despite the fact that its direct effects were so subtle as to be barely measurable, is when it influences the identity of the eventual dominant in a succession by causing a slight change in the competitive ability of one of the species struggling for possession of limited resources (e.g. a transgenic organism that reduced the effectiveness of the way in which a tree species exploited its mycorrhizal fungi, might mean that species B eventually dominates the community rather than species A).

GENETICS AND GMOs

The focus of this paper is ecology, not genetics, but a number of points should be made in passing. In those cases where the introduction of GMOs does lead to the establishment of a self-replacing population, then evolution of the GMOs will certainly occur. Will the transgenic trait increase in frequency within the population or will it decline? The transgene can persist, of course, without any conventional evolution (it might be neutral in fitness terms, or it might be protected from natural selection in dormant stages like long-term seed banks in soil). The Red Queen hypothesis argues that species need to run as fast as they can, just to keep up, and that there is no evolutionary state of perfection which, once reached, allows the species to relax.

"Now, here, you see, it takes all the running you can do, to keep in the same place. If you want to get somewhere else, you must run at least twice as fast as that."
(Through the Looking-Glass, Chapter 2, Lewis Carroll).

While on the subject of Alice and the Queen, it is worth noting what the Queen had to say about ecological risk assessment for the introduction of GMOs:

"There's no use trying" [Alice] said: "one can't believe impossible things.
"I daresay you haven't had much practice" said the Queen. "When I was your age, I always did it for half-an-hour a day. Why, sometimes I've believed as many as six impossible things before breakfast".
(Through the Looking-Glass, Chapter 5, Lewis Carroll).

SO WHAT??

It is important for ecologists to define clearly what it is that they can, and can not, contribute to the debate about what kind of GMOs should be introduced into the environment. How do we separate the trivial from the problematical, the actual from the possible ? What ecosystem functions (if any) must we not interfere with? The answers to these questions will be obtained only with experience from carefully designed experiments, aimed at quantifying the relevant parameter values under a range of field conditions. As I have pointed out already, theory alone will not be enough.

In the clear light of hindsight, a good many of the steps in the step-by-step approach to risk assessment might appear to be unnecessarily cautious, or even completely unnecessary. But this does not undermine the process. What is required is an experimental protocol that would be capable of measuring the percentage difference in ecological performance between a transgenic organism and its conventional counterpart, if such a difference were to exist. Ideally, the protocol should also be able to show why any difference in performance came about (e.g. through altered competitive ability, herbivore tolerance, growth rate, seed production, or anti-predator behaviour). Next we need to be able to predict how the change in performance measured in small-scale experiments will map to an ecosystem-level impact, once the release has been scaled up to the kind of spatial and temporal extent that would be involved in a full commercial introduction. Both of these steps require substantial new research.

The reason that governments treat GMOs differently from other kinds of introduced organisms (like biocontrol agents or new garden plants) is because of fear and uncertainty. Most of the fears are completely unjustified, but not all. There is a good deal of uncertainty about the consequences of introducing GMOs, but we live in an uncertain world. While we shall never be able to predict *exactly* what will happen at an ecosystem level, we can predict certain outcomes within reasonably narrow bounds, so long as we have sufficient empirical information about the organism and the environment into which it is to be introduced. For example, we can predict the weather in England at noon on a given day in January next year, to the extent that we can be reasonably confident that the temperature will lie between -5 and +10 ^{o}C. The worst way to deal with fear and uncertainly is by denial (e.g. by saying that GMOs are just like any other organisms, so there is nothing to be concerned about). This is an example of what we might call the "Trust me. I'm a doctor" syndrome. The time is not far off when introductions of GMOs will be proposed where we have no direct experience, and where small-scale, contained experiments are impossible. Suppose, for example, that someone decided to extend the northward range of an important marine fish species by introducing antifreeze genes. You could not do contained experiments on the ecological impact of such an introduction because of the scale of containment; the ecosystem-level properties in which you would be interested simply can not be reconstructed within experimental enclosures. So what would you do? It seems to me that you only have two choices. You could say "It's only a fish, and fish are harmless, so go ahead". Or you could say "There are serious uncertainties, and the consequences of a mistake could be great, so prohibit the release". Unless we have a well-developed body ecological information, in the form of carefully controlled and closely monitored experiments, built up by addressing increasingly difficult questions in a step-wise manner, then it is going to be impossible to convince sceptics that the introduction of novel GMOs is worth the risk.

REFERENCES

Andrewartha, H.G. & Birch, L.C. (1954). The Distribution and Abundance of Animals. University of Chicago Press: Chicago.

Bazzaz, F.A. (1979). The physiological ecology of plant succession. Annual Review of Ecology and Systematics, 10, 351-371.

Begon, M., Harper, J.L., & Townsend, C.R. (1990). Ecology: Individuals, Populations and Communities. Blackwell Scientific Publications: Oxford.

Bender, E.A., Case, T.J., & Gilpin, M.E. (1984). Perturbation experiments in community ecology: theory and practice. Ecology 65: 1 - 13.

Boucher, D.H., James, S., & Kesler, K. (1984). The ecology of mutualism. Annual Review of Ecology and Systematics, 13, 315-347.

Brown, J.H. (1986). Experimental community ecology: the desert granivore system. In J. Diamond & T. J. Case (Eds.), Community Ecology (pp 41-61). Harper & Row, New York.

Brown, J.S. & Venable, D.L. (1986). Evolutionary ecology of seed-bank annuals in temporally varying environments. American Naturalist 127: 31 47.

Chesson, P.L. (1986). Environmental variation and the coexistence of species. In J. Diamond & T.J. Case (Eds.), Community Ecology (pp. 240-256). Harper & Row: New York.

Clements, F.E. (1928). Plant succession and Indicators. H.W. Wilson: New York.

Colinvaux, P.A. (1973). Introduction to Ecology. New York: John Wiley.

Connell, J.H. (1979). Tropical rain forests and coral reefs as open non-equilibrium systems. In R.M. Anderson, B.D. Turner, & L.R. Taylor (Eds.), Population Dynamics (pp. 141-163). Oxford: Blackwell Scientific Publications.

Connell, J.H. & Slatyer, R.O. (1977). Mechanisms of succession in natural communities and their role in community stability and organization. American Naturalist 111: 1119 - 1144.

Crawley, M.J. (1986). Plant Ecology. Oxford: Blackwell Scientific Publications.

Crawley, M.J. (1992). Natural Enemies: The Population Biology of Predators, Parasites and Diseases. Blackwell Scientific Publications: Oxford.

Crawley, M.J. (1992). Seed predators and plant population dynamics. In M. Fenner (Ed.), Seeds: The Ecology of Regeneration in Plant Communities (pp. 157-191). CAB International: Wallingford.

Elton, C.S. (1958). The Ecology of Invasions by Animals and Plants. Methuen and Co.: London.

Gleason, H.A. (1926). The individualistic concept of plant succession. Bulletin of the Torrey Botanical Club, 53, 7-26.

Grime, J.P. (1979). Plant Strategies and Vegetation Processes. John Wiley Chichester.

Grubb, P.J. (1986). The ecology of establishment. In A.D. Bradshaw, D.A. Goode, & E. Thorp (Eds.), Ecology and Design in Landscape (pp. 83-97). Blackwell Scientific Publications: Oxford.

Hairston, N.G. (1989). Ecological Experiments: Purpose, Design, and Execution. Cambridge University Press: Cambridge.

MacArthur, R.H. (1972). Geographical Ecology. Harper & Row: New York.

Mack, R.N. (1981). Invasion of Bromus tectorum L. into western North America: an ecological chronicle. Agro-Ecosystems, 7, 145-165.

May, R.M. (1973). Model Ecosystems. Princeton University Press: Princeton.

Maynard Smith, J. (1982). Evolution and the Theory of Games. Cambridge University Press: Cambridge.

Maynard Smith, J. (1974). Models in Ecology. Cambridge University Press: Cambridge.

Oksanen, L., Fretwell, S.D., Arluda, J., & Niemela, P. (1981). Exploitation ecosystems in gradients of primary productivity. American Naturalist, 118, 240-261.

Pielou, E.C. (1969). An Introduction to Mathematical Ecology. Wiley Interscience New York.

Roughgarden, J. (1986). A comparison of food-limited and space-limited animal competition communities. In J. Diamond & T.J. Case (Eds.), Community Ecology (pp. 492-516). Harper & Row: New York.

Skellam, J.G. (1951). Random dispersal in theoretical populations. Biometrika, 38, 196-218.

Vitousek, P.M. & Walker, L.R. (1989). Biological invasion by Myrica faya in Hawaii: plant demography, nitrogen fixation, and ecosystem effects. Ecological Monographs, 59, 247-265.

Williamson, M. (1993). Invaders, weeds and the risk from genetically manipulated organisms. Experientia, 49, 219-224.

PLENARY DISCUSSION

Question:
I would like to ask dr. Crawley about two principles: One is about the concept of familiarity with the receptor-organism and its importance to the assessment of its possible risk; the other is about the familiarity with a trait having a selective advantage in the case of relevant selection pressure. Would you like to comment on this?

Response:
Clearly, once you have introduced GMOs they will evolve. And I think the discussion on the ways GMOs evolve following release is a major topic, perhaps for another day. You mentioned traits and predictability or familiarity. From an ecological point of view traits are hopeless for prediction because we are not trying to predict simple things like will it express a morphological phenotype or will it express something like herbicide tolerance -we assume that it will. The things we are trying to predict are complex phenotypic traits like net reproductive rates, will it increase or decrease, will it enhance or reduce the competitive ability of the organism it is linked with. It is not 'traits' as such that are likely to be important in those circumstances, but it is parameter values. So, the distinction I want you to take away is that we are less interested in the effects of GMOs on traits of the phenotype and much more interested in the effects of GMOs on the parameter values exhibited by groups of GMOs.

Question:
I might add to this that after having had discussions with many ecologists I always get this hopeless idea: We will never manage. One of your statements was that ecological generalizations are useless for predictions. What most of us are in fact doing in their daily work, however, is trying to predict to the best of our knowledge what might happen and to make decisions based on it. Could you please comment on this?

Response:
I am not at all as negative as you might imagine from some of the things I say! However, since the effects we are talking about also depend on the weather, I could better have been a private consulting meteorologist than an ecologist. It is quite unreasonable to assume that ecologists can predict the weather and then everything else. What you should ask ecologists to do is to predict things that they would be reasonably expected to be competent to predict. For example, what is the likely fate -relatively speaking- of a transgenic variety of a particular GMO versus a conventional variety. Ecologists could do that. That is the kind of prediction that they could make, but this is not the kind of prediction typically asked to make. Ecologists are often asked to do things that are literally impossible.

Question:
I was very much impressed by the presentation of dr. Crawley. I have one question. I wondered whether you have left out of your presentation on purpose the

intraspecific genetic variability which can lead to very unusual developments, may be even to unpredictable developments?

Response:
I think I did refer to it in passing. The example of Salvinia floating fern was as predictable as it was because there was no intraspecific genetic variation at all (the species is a sterile tetraploid). So, I left it to you to extrapolate from that; if there was intraspecific genetic variation, then life would be more difficult.

Question:
You said as ecologist you could predict the survival, invasiveness or whatever of a GMO compared to a traditional species. My question is, if ecologists could indeed do that, how much time of study would he need. Could he do it and provide the people who have to decide on regulations with an answer within a week, the next month or year, after a decade?

Response:
This is clearly an important question. And I have some personal experience; it takes longer to get permission to do it, than it does to do the work! When I said ecologists could make these kind of predictions, it would be based on a thorough and well focused body of empirical evidence that bore specifically on that question. The whole point is that theory is useless because it can go one way or the other, depending on the parameter values. I showed you the behaviour of an one parameter model of population dynamics. Nothing could conceivably be more simple and yet that varied from rock solid stability to totally unpredictable chaotic dynamics, on the minor tweaking of a single parameter. Without empirical knowledge of the parameter value, that is likely to be influenced by GMOs, you simply cannot make the prediction. You must have empirical evidence, you cannot do it in theory. What you can do, though, is build up experience. It would be absurd to argue, and I certainly would not do that, that part of the step-by-step and case-by-case approach is that you have to do field experiments of the scale and ambitiousness of the PROSAMO-project for every intended introduction. That would be ridiculous. What you can do is to begin to argue things like this: We now know a certain amount about transgenic oilseed rape which expresses a herbicide tolerant transgenic phenotype, and perhaps we can begin to think about relaxing the constraints involved in the permission to release. These are the kind of discussions going on now. But those discussions are now better informed from an ecological point of view than they were before. To that extent the discussion is a more sensible one. But you are talking about years rather than weeks of work for any particular class of organisms. If for example you are now to ask what would happen if we start considering transgenic perennial forage crops, then these would have a very different ecology from something like oilseed rape which would only grow naturally in arable fields. They are much longer lived, and much more likely to be invasive. If you envisage introducing very different kinds of constructs in very different kinds of crops, like for instance a drought tolerant perennial rye grass, then you would need to do several years of work, if you regarded the risks as justifying the effort.

Intervention:
You have given us a very short and very helpful definition of what we should mean with long term. You have told us that this means approximately one hundred years. In my view, it becomes obvious that we do not have any experience with such a long term ecological behaviour of transgenic organisms. Therefore, we must look for a kind of indirect approach. That means that we have to look for, say, ecological models; models that can be powerful instruments in modelling the ecological behaviour of the GMOs. I think, we all know two different models. The first one is to take the original non-transgenic plant as a kind of ecological model for the behaviour of the transgenic plant. The other one is known as the Exotic Species model of the naturalization of alien plants from the point of view of invasion-biology. Both models are very useful instruments to predict particular items of the ecological behaviour of transgenic organisms.

Response:
There are at least two quite long seminars that could be organized on the issue of modelling; the first one on the use of non-transgenics as a model for the behaviour of transgenics, the other seminar on the Exotic Species Model. If you look at the Introduced Species Model and you look a the data, one of the predictions is that different kinds of communities differ in invasibility. Some communities are more invasible than others. However, these kind of generalizations just simply do not stand up, though they are statistically sound. You can look at what plant families are more invasive. There are things like Crucifereae are more invasive and Orchidaceae are less invasive. But if you got two Crucifers side by side, you cannot tell which is going to be the more invasive because you do not have the parameter values.

Question:
I enjoyed your presentation and I think a learned a little more ecology. I would like to ask a question on the predictions and the models that you displayed. Do those also hold for agricultural crop species, on which most of the decisions are being made. In many of those the survival is actually being manipulated by man. And many of those are also introduced species. Could you comment on that?

Response:
There is a big issue here about the extent to which agricultural crops confined within the field boundary could ever be considered to be a problem that an ecologist as a risk assessor would want to address himself to. There is a school of thought that if there are problems, these are farmer's problems (like for example volunteer weeds and the risk of having herbicide resistance in volunteer weeds.) The thing where an ecologist would come in would be to look at a slightly longer term. To take an agricultural example: What if your genetic modification caused your oilseed rape seeds to have protracted dormancy. Unintentionally you increase the dormancy, you then increase the duration of time over which volunteer weeds will come up from the seedbank. That is the kind of issue that an ecologist might usefully address.

TOPIC 1

INSECT RESISTANT PLANTS

INTRODUCTION INTO THE BACILLUS THURINGIENSIS ISSUE

Prof. dr. S. Tolin (rapporteur), Department of Plant Pathology, Physiology & Weed Science, Virginia Polytechnic Institute & State University, United States of America

The bacterium *Bacillus thuringiengensis* is a pathogen of certain healthy, uncompromised insects. Preparations of *B. thuringiensis* have been used commercially as microbial insecticides for about 25 years worldwide, with isolates of *B. thuringiensis* subspecies *kurstaki* most widely used for agronomic, silvicultural, and horticultural applications. Its record of safety is, in part, attributed to its biological specificity. It is pathogenic only to leptidopteran larvae having a high gut pH, and has no effect on humans, other vertebrates, or nontarget organisms.

The efficacy of *B. thuringiensis* as a biopesticide is limited by its lack of persistence, thus repeated applications to foliage are needed. The bacterium is UV inactivated and is washed from leaves by rainfall. Its rate of germination in soil has been described as "quite sluggish". Even with repeated applications, it apparently does not proliferate to epidemic numbers and is not widely disseminated from the site of introduction. The ecological niche of *B. thuringiensis* is unique, in that insect larval haemolymph provides an excellent nutrient environment. Even through several other microorganisms occur in the haemolymph, there is no evidence of transfer of sequences of the toxin gene to other species.

Two practical considerations have also limited more extensive use of *B. thuringiensis*. The first is that microbial pesticides are much slower-acting than are chemicals. For this reason some commercial formulations of *B. thuringiensis* at times contain chemical insecticides to decrease response time and satisfy users. The second is the inevitable problem of larvae developing resistance to the toxin, which was first recognized in 1985 in insects following extensive use of "Bt" in grain storage bins.

The mechanism of pathogenicity of *B. thuringiensis* is well-understood. Following ingestion by feeding larvae, the bacterium proliferates and may sporulate in the nutritious haemolymph. Spores, if formed, contain a number of paracrystalline or crystalline inclusions, composed of proteinaceous protoxins. Upon ingestion of such inclusion-containing spores by susceptible larvae, (i.e. Leptidoptera and Diptera), the protoxin is solubilized in the alkaline, reducing conditions of the midgut. Proteases in the midgut also process the protoxins to toxic protein(s), among which ∂-endotoxin has been reported to have the greatest specificity and activity. Specific receptors in the midgut epithelium are triggered, the gut and mouthparts are paralysed, and the larvae stop feeding. Further severe pathogenic effects, both physiological and cytological, have been documented in the infected insects.

Genes encoding δ-endotoxins and other toxins have been cloned and sequenced from various isolates of *B. thuringiensis*, and factors relative to their expression and regulation are known. Depending on the isolate, both chromosomal and plasmid DNA sequences encode the toxin(s) or regulatory factors. Numerous improved strains, including those whose toxin expression has been enhanced by genetic manipulation, including by recombinant DNA methods, are in commercial use in many countries. A great deal of familiarity with this system exist.

Since lepidopteran insects susceptible to *B. thuringiensis* are important plant pests, a logical question for an early plant biotechnology experiment was to ask whether plants engineered to contain the gene for one of the δ-endotoxin in plant cells, including from the total gene or from truncated constructs, caused insect larvae to cease feeding on leaves. These dramatic results spurred the interest of plant biotechnology companies worldwide, and gave the promise of a commercial product with a more effective delivery system for the active compound of a proven system for insect management.

The first field tests of Bt-tomato were conducted in the United States in 1988 by three different companies: Monsanto, Sandoz and Agrigenetics. A fourth test was by Crop Genetics International of the δ-endotoxin-expressing endophyte, *Clavibacter xyli*, with corn. In 1989 field tests of tobacco (Rohn & Haas), cotton (Agracetus, Monsanto), and tomato (Calgene, Monsanto) were conducted. In 1990, Ciba Geigy entered the field with tests of Bt-tobacco, federal agricultural researchers tested Bt-potato, and previous testers continued with advanced generations or with new, improved transformed plants.

In the next three years, advanced scale field tests conducted throughout the United States and worldwide demonstrated that plants providing commercially acceptable levels of tolerance to targeted lepidopteran insects can be obtained. The performance of the plants relative to other traits is now being evaluated, and germplasm for producing commercial varieties, ultimately, is being selected.

Approval for the use of δ-endotoxin in commercial plants for agriculture is being sought. Issues being identified and considered relative to the safety of the use of such plants include those need to establish the food, feed and environmental safety of these plants and plant products. The key concerns to be addressed relative to the potential, long-term and ecological impacts include: (i) increased competitive advantage of wild plants to which the trait might transfer and (ii) increased selection pressure for resistance in insect populations conditioned by potential widespread use of identical toxin(s) in crop plants.

INSECT TOLERANCE IN PLANTS

Dr. M. Peferoen and dr. P. Rüdelsheim, Plant Genetic Systems, Belgium

SUMMARY

Agricultural productivity is to a large extent dependent on the use of pesticides. However, concerns are being raised about their effects on the environment, wild life and on people. Insect tolerant crops can reduce pesticide usage while quality and productivity increase. Insecticidal proteins from plants and from *Bacillus thuringiensis* (Bt) have been introduced in crops such as potato, tomato, cabbages, cotton, corn, rice, apples, walnuts. By the end of this century, several insect tolerant crops will be commercialized. The safety of such crops for the consumer and for the environment should be assessed. There are three main points of concern, the safety for non-target organisms, the development of resistance in insect populations and the ecological impact. Ideally, the activity of the insecticidal ingredients which are introduced into the crops has to be restricted to the target insect pests only. Side effects upon non-target organisms, including human beings, have to be avoided. In general *Bt* proteins have been shown to be very safe for non-target organisms. Potential side effects by novel insecticidal proteins should be determined in toxicological evaluations. Also, the insect resistant crops should be designed and deployed in such a way that resistance development in insect populations is retarded. A number of pest resistance management tactics can be incorporated in the crop itself. In addition, such crops can also be used in integrated pest management programs. Invasiveness of crops is determined by a range of characteristics and is highly variable between crops. Insect resistance may increase the potential of crops to colonize natural habitats. However, crops with only a few weedy characteristics are unlikely to become weeds when supplemented with resistance to certain insects. Some of the ecological aspects can be tested in field trials, but most can be deduced from the extensive experience gained with crops obtained through classical breeding. Finally, risk assessment of insect tolerant crops should be part of a risk - benefit assessment of crop protection in general.

INTRODUCTION

Each year farmers spend in excess of U.S. $23 billion on pesticides, including U.S. $6 billion on insecticides. While the benefits of pesticides are acknowledged, concerns have been raised about their effect on the environment, wild life and on people. Despite these concerns, increased agricultural productivity has relied heavily on the use of pesticides. Indeed, pest control is a key factor influencing productivity. Genetically engineered insect tolerant crops can become an important environmentally friendly means of insect control and help aid agricultural production.

The safety of insect tolerant crops should be assessed, including the safety of the active ingredients for non-target organisms. That assessment should take place within the context of a risk-benefit assessment. Since insect populations can adapt to different control measures, strategies to retard or avoid development of resistance should also be studied. Finally, the ecological impact of weediness developing amongst insect resistant crops, or of their hybridizing with related weeds, should be assessed.

INSECTICIDAL PROTEINS

Insect tolerance in crops is due to the presence of insecticidal proteins, encoded by single genes. There are two main sources for insecticidal proteins: plants and the bacterium *Bacillus thuringiensis.*

Plant derived proteins

During their evolution plants have developed mechanisms to protect themselves against insects. In general, plant-derived insecticidal proteins can be classified as enzyme inhibitors, such as protease and amylase inhibitors, lectins and ribosome inactivating proteins. Some protease inhibitors inhibit serine proteases present in the digestive tract of Lepidoptera, while others block thiol proteases in the gut of Coleoptera. The cowpea trypsin inhibitor, a protease inhibitor, blocks protein digestion in insects, while alfa-amylase inhibitor disrupts carbohydrate metabolism. Lectins such as wheat germ agglutinin and snowdrop agglutinin are sugar binding proteins whose insecticidal efficacy is due to their binding to certain compounds in the gut epithelium. The ribosome inactivating proteins (RIPs) are N-glycosidases which remove an adenine base from the ribosomal RNA of the 28S eukaryotic ribosome.

Some plant-derived insecticidal proteins are potentially toxic to non-target organisms. For example, wheat germ agglutinin has antinutritional effects in rats, and most RIPs are more active on mammalian ribosomes than on ribosomes of other eukaryotic organisms. The effects of plant derived proteins on non-target organisms should therefore be assessed.

Bacillus thuringiensis

Bacillus thuringiensis is a sporulating bacterium which can be applied as a spray to protect crops from insect attacks. Today *Bacillus thuringiensis* (Bt) is the most successful biopesticide with annual sales ranging between U.S. $60 and $100 million. Used as a biological insecticide for over 30 years, Bt has proven to be very safe.

The insecticidal activity of Bt is determined by proteins which are produced during sporulation and are stored as crystals. Today 25 crystal protein genes have been cloned. Crystal proteins have been found to be toxic to lepidopteran, coleopteran and dipteran larvae. Each crystal protein is characterized by a very specific spectrum of activity which is limited to a few insect species. Numerous studies with different Bt preparations on a wide range of non-target organisms have confirmed its extreme specificity.

The mechanism of action of Bt is well known. Upon ingestion, the crystal protein is proteolitically activated to release the toxin, which binds to receptors in the insect midgut. After binding to the receptor, the Bt toxin, either directly or indirectly, disrupts the cell membrane leading to colloid osmotic lysis of the epithelial cells, gut disintegration and eventually to the death of the insect. The extreme specificity of the different crystal proteins is to a large extent determined by the presence of receptors in the insect midgut. It has also been demonstrated that there are different Bt receptors for different Bt proteins, even within the same insect.

For plant genetic engineering, the Bt protein is expressed in a truncated form. The part removed from the protein is thought to be involved in the formation of the crystal in the bacterium, not in the activity. Some strains which are used in bioinsecticide sprays produce naturally truncated crystal proteins and have no effects upon non-target organisms. Experiments whereby protoxins are truncated have demonstrated that truncated proteins have the same activity and spectrum as the protoxin. Toxicological studies with a truncated Bt protein expressed in *E. coli* has demonstrated its safety up to 4000 times the maximum concentration man is likely to consume. Each crystal protein has its own activity spectrum and results obtained with one crystal protein cannot readily be transferred to other crystal proteins. Use of a new crystal protein may require some specific toxicological evaluations.

INSECT TOLERANT CROPS

High levels of cowpea trypsin inhibitor introduced in tobacco protects the plant from damage caused by tobacco budworms. Wheat germ agglutinin has been introduced in corn and experiments are under way to introduce a snowdrop lectin in rice to protect it from attacks by brown planthopper and green leafhopper.

Since 1987 numerous crops have been engineered with different crystal protein genes. Genes from the *cryIA* group have been introduced into tomato, potato, cotton, corn, rice, apple and walnut to control lepidopteran pests. Potatoes have been engineered with the *cryIIIA* to control Colorado potato beetles. Several crops containing a Bt crystal protein gene are being evaluated in field trials in the USA, Europe and China.

RESISTANCE

Resistance to insecticides has been documented in more than 500 insect species. When insect populations show resistance to an insecticide, the tendency has been to increase the treatment regime. This leads to a vicious circle of increased incidence of resistance and increased use of the insecticide, until finally the insecticide has virtually no effect even at an extremely high dosage. As a consequence pressure on the environment increases, crop yield decreases, and insect-borne diseases spread. When all control measures have been used and exhausted, there is nothing left but to surrender to the insects. This has happened on several occasions, and is, for example, why Thailand imports cotton while it was once a cotton exporting country.

The development of resistance depends on the genetics of the resistance mechanism, on the selection pressure caused by the insecticide, and on the dynamics of the insect population. A low resistance-gene frequency in the population and an associated fitness cost will retard the development of resistance. Similarly, a regular influx of susceptible genes in a population under pressure off-sets resistance development. Continuous exposure to a single insecticide through the insect's development is highly conducive to resistance development.

So far only one insect species, diamondback moth (*Plutella xylostella*), has developed resistance in the field to Bt sprays. Diamondback moth is notorious for its potential to develop resistance against insecticides. When treatment of diamondback moth with synthetic insecticides became ineffective, farmers began exclusively using Bt and thereby rapidly eroded its use.

Analysis of the resistance mechanism has indicated that it is based on a change in one of the insects Bt receptors. Interestingly, because other Bt receptors are still accessible, the resistant diamondback moth populations can still be controlled using other Bt proteins. Also, the resistance appears to be recessive and comes with a fitness cost, which may contribute to the loss of resistance when the population is no longer exposed to Bt.

Resistance management tactics aim to retard resistance development by reducing the selection pressure on the insect population. It is often assumed that a constitutive expression of a single insecticidal protein in a crop will inevitably lead to the development of resistance. However, if the target pest feeds on other crops and weeds which do not contain the insecticidal protein, then the pressure on the entire population is actually quite low and resistance may never develop. Also, the insect resistant crop can be part of an integrated pest management system, resulting in a balanced selection pressure.

Transgenic plants in fact present a unique opportunity to apply pest resistance management (PRM) tactics. For example, expression of the insecticidal protein can be targeted to only particular parts of the plant, leaving other plant parts as refuges for susceptible insects. Because the level of insecticidal proteins expressed by a plant is more homogeneous than an insecticide sprayed over the entire crop, insect resistant crops can be used in a high dose strategy. Expression of multiple Bt proteins at a high level — killing all but the homogeneous resistant insects — in combination with a refuge which promotes dilution of the resistant with susceptible insects, is considered a promising strategy.

ECOLOGICAL IMPACT

Ecological concern with genetically engineered insect tolerant crops is focused on 1) whether insect resistant crops will have an increased invasiveness which allow such plants to colonize natural vegetation and 2) the possibility that the insect tolerance trait spreads to wild and other relatives of the engineered plant.

While insect resistance can make crops more competitive, the question is whether this will be enough to push the net reproductive rate above 1, leading to invasions. Several factors determine a plant's invasiveness, including its growth rate, seed production, and competition capacity. Little is known concerning traits which enhance invasiveness. Invasiveness and hybridization potential of genetically engineered crops can be assessed in field trials. Because field trials are mainly small scale, they cannot easily assess the ecological impact of the large scale cultivation of such crops.

Experience with crops obtained through classical breeding can be used as a basis for the safety assessment of engineered varieties of such crops. None of the crops which have been classically bred for insect resistance by crossing them with wild relatives have been reported to cause serious ecological problems.

Crops which in some environments are regarded as weeds, such as sunflower, bermuda grass and rice, would require an ecological assessment when engineered with a disease or pest resistance. Other crops such as corn, wheat, tomato, soybean have very few "weedy" characteristics and protection from a few insect species, is unlikely to increase the invasiveness. Although certain traits influence the invasiveness potential, it is the overall characteristics of the crop which need to be taken into account in order to assess its potential ecological impact.

CONCLUSION

There are no universal guidelines for the deployment of insect tolerant crops. Any such evaluations will have to proceed on a case by case basis. Cabbage expressing a *Bacillus thuringiensis* protein to control diamondback moth larvae, is very different from a potato plant expressing wheat germ agglutinin to control aphids. For example, the toxicology of the insecticidal proteins is different, the two pests feed on different crops, cabbage and potato vary in weediness, finally the hybridization potential with wild relatives is relatively low in potato compared to cabbage.

For each crop, the safety of the insecticidal ingredient, the effect upon the targeted insect populations and the ecological impact has to be assessed. To manage resistance development in the insect populations a number of tactics are available to retard the development of resistance. Implementation of such tactics will require collaboration between technology providers, seed companies, regulatory agencies, extension workers and farmers. To assess the ecological impact of insect tolerant crops, substantial information is available from extensive experience with classically bred crops. Finally, any risk assessment should be part of an risk-benefit assessment. The benefits of insect tolerant crops for sustainable agricultural production should be considered.

REFERENCES

Hilder, V.A., Brough, C., Gatehouse, A.M.R., Gatehouse, L.N., Powell, K.S. & Shi, Y. (1992) Brighton Crop Protection Conference, Proceedings Vol 2, 731-740.

Lambert, B. & Peferoen, M. (1992) BioScience 42, 112-122.

Koziel, M.G., Beland, G.L., Bowman, C., Carozzi, N.B., Crenshaw, R., Crossland, L., Dawson, J., Desai, N. Hill, M., Kadwell, S., Launis, K., Lewis, K., Maddox, D., McPherson, K., Meghji, M.R., Merlin, E., Rhodes, R., Warren, G.W., Wright, M. & Evola, S.V. (1993), Bio/Technology 11, 194-200.

Fujimoto, H., Itoh, K., Yamamoto, M., Kyozuka, J. & Shimamoto, K. (1993) Bio/Technology 11, 1151-1155.

Van Rie, J. (1991) Tibtech 9, 177-179.

McGaughey, W.H. & Whalon, M.E. (1992) Science 258, 1451-1455.

Keeler, K.H. (1989) Bio/Technology 7, 1134-1139.

Crawley, M.J., Hails, R.S., Rees, M., Kohn, D. & Buxton, J. (1993) Nature 363, 620-623.

Raybold, A.F. & Gray, A.J. (1993) J. Appl. Ecol. 30, 199-219.

PLENARY DISCUSSION

Questions:
The question I have is what do you mean by case by case? Is that toxin per toxin? Is that construct per construct? Is that transformant per transformant? Secondly, I am very pleased to see that industry takes a very pro-active role. Let us assume resistance will develop, what are we going to do about it? One of the elements with relevance to this discussion is: Is it a single gene or multiple gene? Or another element: Is the promoter is a constitutive one or is it a stress induced one, or a tissue specific one? Would you agree, in order to avoid these kind of resistances, with the simple conclusion that single genes with the use of a constitutive promoter like CaMV 35S promoter should not be allowed?

Response:
Your first question is very difficult to answer at this stage, because the level of experience is rather limited. I can quote an example: We were trying to assess the toxicology of one Bt protein expressed in a tomato plant, because we think that the interaction between a Bt protein and a certain genetic background could be different from another genetic background. We did do these toxicological assays. The problems is toxicologists saying: We need to apply more of the protein because if you look at the expression levels, especially those of the Bt gene, these are fairly low. So, if you try to incorporate that into traditional toxicological assays, you have a big problem. Our rats were dying from the control tomatoes they were eating. So, we could not really feed them enough Bt containing tomatoes. That is why they took a second approach, which is to try as to feed as much as you can the real protein that is in the plant. At this stage it is very difficult to speculate whether a Bt protein in a potato and the same Bt protein in tomato will differ in a toxicological sense. I think this may require some additional toxicological testing.
Concerning your second question, again this is a difficult one. It depends on, for example, the population size you are targeting. If you are planning to introduce this one specific Bt protein in a wide range of crops on a massive scale, then you are probably in a very dangerous situation. If you look at cotton where you introduce this single Bt protein under a constitutive promoter: Many of the cotton pests live on other wildtypes, on weeds and on other crops. So, it is very well possible that, because of the size of your plots in relation to the total area where these insects feed on, it does not really make any difference because there is a continuous flow of genes around that is diluting the resistance development. There a few scenarios where this very blunt approach still could work. That depends on the ecology of the insect. If the insects stays on the same plant as almost an exclusive feeder, then you are in a dangerous situation.

Question:
In your pro-active description you did not mention the obvious possibility of combining this Bt protection with insecticide protection which would have the advantage that farmers could continue to use a product that they have been using before but that they could no longer use because of resistance development. If they could use Bt crops, they could go back to cheap, maybe better products, because the resistance would be handled by the Bt and vice versa. Did you have a reason not to mention this?

Response:
No, I just put this issue in the whole bag of the 'Integrated Pest Management' approach, whereby indeed, by having these plants available you become a lot more flexibility in your approaches. Bt plants are certainly not the only solution to insect problems. They are just another tool and they may open up additional tools; tools we may have lost in the past and which may become available again.

Question:
We are talking a lot about the development of resistance among plant pests. But what do we know about the role of the Bt protein about regulating natural insect populations in natural environments?

Response:
Epizootics of Bt are very rare. You only find epizootics of Bt in contained environments like, for example, grain bins where all of a sudden you have a complete breakdown of your insect population; it turns out to be infected by Bt. In forests, in agricultural systems, and in weedy situations you never find that. Bt is actually a very poor pathogen. Actually it is not really sure that Bt is producing insecticidal crystal proteins to kill insects; maybe it has another reason to do that. So, the role of Bt controlling insect populations in natural situation is probably very limited.

Question:
Insects have several genomes in them; there is the insect genome itself but there are also microbial symbionts, of which there is increasing evidence that they play various important roles in insect physiology. Is there any evidence for resistance mechanisms to insecticides (there is probably not enough information on Bt yet) being due to symbiont-encoded properties. And if so, could one envisage a long-term scenario of horizontal spread of these forms of resistance?

Response:
That is a tough one. For Bt no; all scenarios go back to this receptor modification. There are a few things coming up in laboratory selection where the receptor modification scenarios do not really fit in, where probably some other mechanisms are involved. About the role of symbionts; even in the case of synthetic insecticides, I do not know.

Question:
Has anybody ever looked into that?

Response:
Not in all cases of resistance is it clear what the mechanism is. Those that have been published can clearly be traced back to the insect genome. This does not exclude the possibility of horizontal spread just mentioned, but so far I have not seen any report on that.

Question:
Do the Bt genes only work against grazing insects or do they also work against sucking insects? I also have a comment which could be directed to the regulatory authorities. We have some experience in the history of plant breeding which may have relevance for the Bt story. In the theory of plant breeding we talk about resistance breeding using major resistance genes (these are single genes) causing total resistance (so there is a high expression level) but usually the effect is short lived because the pathogen or insect can adapt. And the most pregnant story in this respect is that about the potato late blight. In the twenties plant breeders started to introduce foreign genes into potatoes and they caused a temporary resistance to potato late blight at a very high level. Then one gene after the other failed. Plant breeders responded by accumulating a number of resistance genes and the fungus responded by accumulating equivalent numbers of virulences. In the end, somewhere in the fifties, plant breeders had to give up this approach. We call this generally the Boom and Bust Cycle. And we may hypothesize that with Bt we will get engaged in Boom and Bust Cycles. Then you can say: Well, that is good for the company because they can sell every time another gene. And that is what the plant breeders also said. This brings to a point for the regulatory authorities: For the time being this is all nice and fine but in the long run shouldn't we look for non-monogenetic resistances which are supposed to hold longer and give more stability?; and should we not move away from constitutive promoters to tissue specific promoters? If we could look forward, learning from history, then we could point out a certain direction.

Response:
Most of the Bt proteins do not work against sucking insects but there are a few that do work on them. In relation to your remark, I think it is very difficult to just say that the use of a single gene or a constitutive promoter will lead to resistance development in all cases. Yes, there will be such cases. But there can be other cases where the ecology of the insect is such that you have a constant mix of the population and thereby a dilution of resistance development.

ASSESSING INVASION RISK

Dr. R. Hengeveld, Institute for Forest and Nature Research, The Netherlands

SUMMARY

In this paper, I first analyze the well-known invasion of the Collared Dove, *Streptopelia decaocto*, into Europe to obtain insight in the most important process components determining the rate of invasion. To this end, I use a mathematical model that has recently become available and that has been tested using information on several vertebrate species. Population growth, together with dispersal rate are the main parameters determining the rate of a species' invasion. Of these, long-distance dispersal appears an important and sensitive subcomponent. Information on these parameters has been evaluated for *Bacillus thuringiensis*, showing its inadequacy for invasion risk assessment. Moreover, risk assessment is even more difficult to do for transgenic plant species in which genes from *B. thuringiensis* producing the biopesticidal protein δ-endotoxin have been inserted. Finally, resistance development in herbivorous insects consuming the crop species or variety concerned complicates the possibility to assess potential risks incurred by the manipulation even more. I conclude that risk assessment at the level of the population is too complicated to be applied routinely, even when possible impacts on other species in the ecosystem are left out of consideration. Risk assessment at the level of the individual remains the only practical means for the evaluation of safety measures to be taken.

INTRODUCTION

In this paper, I report on newly available modelling techniques for assessing the risk of invasion of species into a new area. Knowledge of the natural invasion process is necessary for evaluating the potential risk of spreading that novel genetic varieties run, having their new traits obtained by genetic engineering. This spread can occur either through the species of the transgenic variety itself, or horizontally through populations of wild species. Both processes have caused a widely felt concern in society.

I illustrate the application of the new modelling technique by analyzing the best-known invasion so far, that of the geographical range expansion of the Collared Dove, *Streptopelia decaocto* (Friv.) into Europe during this century. Thereby, I emphasize the least-known part of the process determining the rate of spread, that of long-distance dispersal. Using the insight obtained into the determinants of spread as well as into the labour required for parameter estimation in the field, I turn to the potential spreading risk of *Bacillus thuringiensis*, or of the plants into which particular genes from its genome responsible for the production of the insecticidal protein δ-endotoxin have been introduced. Furthermore, I discuss what should be known for evaluating the risk of development and spread of insect resistance to transgenic plants. Finally, I discuss problems of predicting risk in these cases. The new modelling techniques can give a firmer grip on the analysis and quantification of various aspects of risk prediction, structuring this analysis.

It appears that assessing a potential risk incurred by engineering plant or animal species genetically at the level of the population adds a considerable amount of labour and extra cost to the risk assessment already done routinely at the level of the individual. It should be noted that this additional risk assessment does not even evaluate the potential impact of the altered ecological species parameter values at the level of the ecosystem. It seems practical, therefore, to concentrate on risk assessment at the individual level as long as possible, introducing the assessment at higher levels of integration when this seems unavoidable. If, however, this proves too expensive, either in financial terms, or in those of the time and labour involved, one should consider not to introduce the new genetic trait.

RANGE EXPANSION OF THE COLLARED DOVE INTO EUROPE

The range expansion of the Collared Dove into Europe during the present century represents the best worked out example of range expansion in general (cf. Hengeveld, 1988, 1989; Nowak, 1965; 1989; Stresemann & Nowak, 1958; Van den Bosch et al., 1992; and the many references therein). Figure 1 shows that at the turn of the century, this species was still confined to Asia Minor, after which it spread into the Balkan Peninsula where it remained until about 1930. After 1928, it started to spread at a constant rate across Europe (cf. Hengeveld, 1988, 1989; Nowak, 1989).

Figure 1 Spread of the Collared Dove, *Streptopelia decaocto*, into Europe during the 20th century (after Hengeveld, 1989 and Nowak, 1989).

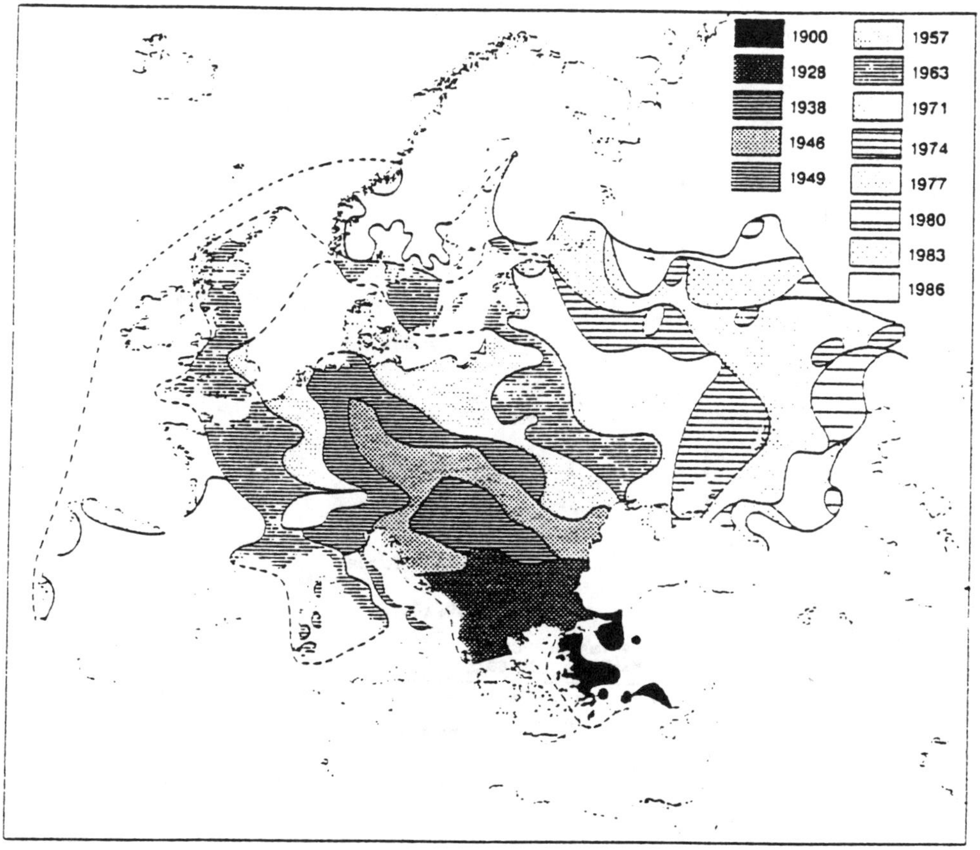

Analysis of the process

Thanks to the great and long-lasting attention paid by amateur and professional biologists alike, Van den Bosch et al. (1992) (cf. also Hengeveld & Van den Bosch, 1992) were able to calculate its expected rate of spread and to compare this calculated rate with the one observed. They used a new mathematical model developed by Van den Bosch et al. (1990) based on a reaction-diffusion process (compare Fisher, 1937; Skellam, 1951), simultaneously applied to several other vertebrate species having expanded their range. In this type of process, the diffusion component represents dispersal and the reaction component population growth after settlement.

Figure 2 Relationship between C_{exp} and C_{obs}. The 45° line representing a one-to-one relationship is drawn independently from the data points (after data in Van den Bosch et al., 1992)

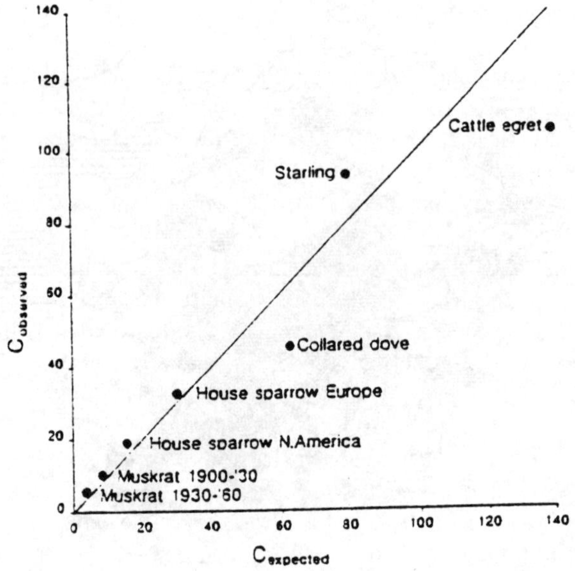

The basic methodology of this model is to separate the explanatory parameters from those characterizing the spread at the observed, geographical level. This methodology, although widespread in physics and chemistry, is new to ecological and biogeographical analysis. It uses data on the species' life history and dispersal capacity at the level of the individual organisms to explain the spread of the population at the geographical level. By not using data from the invasion front itself but from the main part of the range, independence of expectation and observation could be achieved.

The data on population growth concern those of the survival chance of the individuals and the distribution of their fertility over the age classes (Hengeveld & Van den Bosch, 1991; Van den Bosch et al., 1992). These data were obtained from the literature. Multiplying the relative frequency of animals per age class with the fertility rate of this class and summing these products gives the net rate of reproduction, R_o. The dispersal capacity, in turn, was estimated from the distance - or contact - distribution concerning the distances between the nest of a young bird relative to its birth place. Depending on the accuracy required, one can either use the average of this distribution, or add some measures of dispersion around this average, such as the variance, or the kurtosis. Combining these estimates into the equation for the expected rate of spread gives the parameter value for this rate, C_{exp}, to be compared with the observed rate, C_{obs}, estimated from a map (Hengeveld & Van den Bosch, 1991). As these rates should be similar, their values should lie on a 45° line in the space spanned by these two parameters.

Figure 2 shows that the values for the species analysed follow the 45° line reasonably well. It also shows that the dots do not coincide but form an elongated scatter, indicating that the invasion rates differ considerably among these species. Moreover, the data differ for the House Sparrow, *Passer domesticus*, invading North America and Siberia and for the Muskrat, *Ondatra zibethicus*, before and after 1930. These differences imply that the invasion rate is not a specific, intrinsic species property, but that they vary, possibly as a function of environmental conditions. This certainly holds for the Muskrat, not only because the invasion rate lowered because of an intense eradication program running in Germany after 1930, but also because of observed variation in the rate under different conditions. Under unfavourable conditions - dry regions or dry years - it spends fewer animals on building up a local population than on dispersal, resulting in a higher velocity (Schröpfer & Engstfeld, 1983). The invasion rate of the House Sparrow in North America as calculated from European data was half as low as the observed rate. This can be accounted for by the continuous breeding of this species during the initial phase of invasion increasing the net rate of reproduction from 3-4 fledglings per year in Europe to an average of 24 in North America (Barrows, 1898). The same phenomenon, possibly due to social interaction between the males, was also found in New Zealand (Kirk, 1890).

The impact of jump dispersal on spreading rate

It is often observed that bridgeheads occur far ahead of the more closed invasion front. Figure 3 shows this for the Collared Dove. Most young individuals happen to stay and breed in the vicinity of their parent's nest, whereas only a few disperse great distances by long-distance or jump dispersal (Pielou, 1979) (Figure 4), thus forming bridgeheads. After settlement, these bridgeheads can start growing exponentially (Hengeveld, 1988, 1989), similar to the new populations in the invasion front. The latter too expand steadily by short-distance dispersal, sending off the occasional jump dispersers that, in turn, can form new bridgeheads themselves. This composite dispersal process has been called dual dispersal (Zawolek & Zadoks, 1992), and the resulting invasion process hierarchical diffusion (Cliff et al., 1981). A combination of both short- and long-distance dispersal results in the highest spreading rates (Haggett et al., 1977).

The subsequent effects of these two processes can best be seen after the initial introduction of a species into a new continent. Figure 5a shows the initial exponential expansion of the Japanese beetle, *Popilia japonica*, after 1916 in North

America, and Figure 5b shows its rectilinear increase following the exponential increase of the population. It can be imagined that after reaching a certain population size, the absolute number of long-distance
propagules became large enough to start off a sufficient number of independent bridgeheads, and thus to start off the invasion process proper by hierarchical diffusion (Hengeveld, in prep.). This can have been enhanced by the fact that the rate of the linearly increasing perimeter gradually lags behind that of the exponentially growing surface area, resulting in bridgehead formation.

It depends on the ratio of the rate of population growth relative to that of jump dispersal how great the impact of this type of dispersal will actually be; at high growth rates of short-distance dispersers, the bridgeheads can be overrun before having any effect on the expansion velocity.

<u>Figure 3</u> A more detailed map than that of figure 1 of the progress of the spread of the Collared Dove into Europe, showing the bridgeheads ahead of the closed front (after Nowak, 1965).

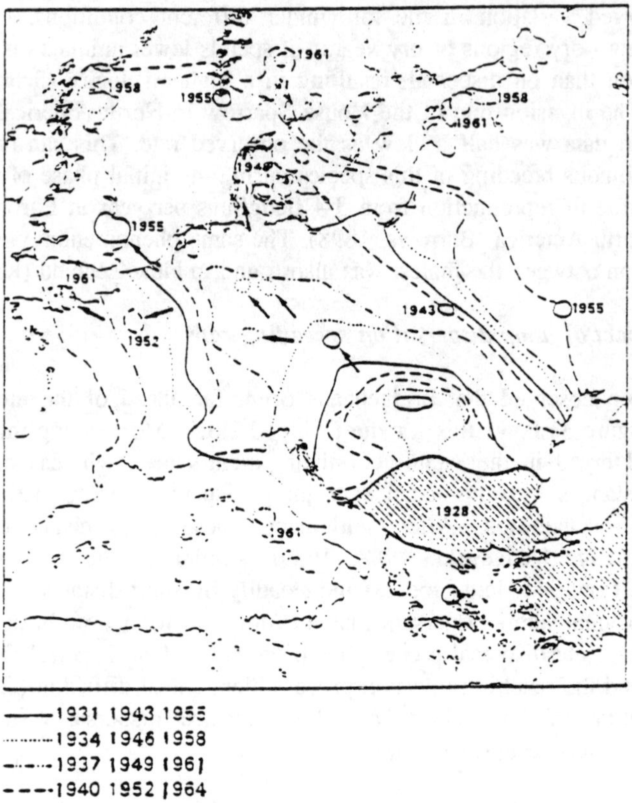

Figure 4 Frequency distribution of distances covered as measured from a map of first sitings of the Collared Dove in Fisher (1953).

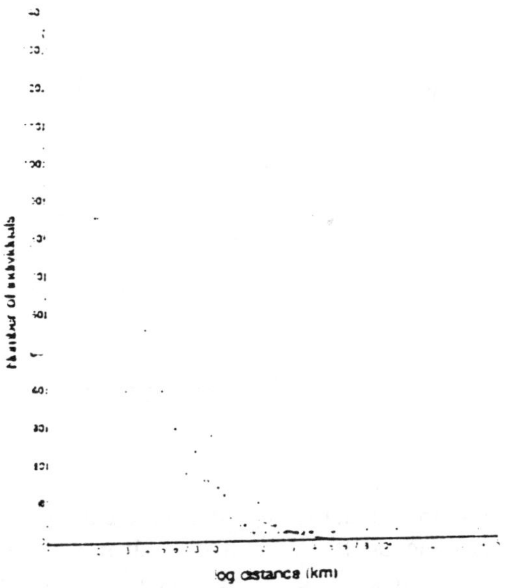

Figure 5a Initial phase of spatial expansion of the Japanese beetles, *Popilia japonica*, into eastern North America during the first part of the 20th century (constructed from a map in Elton, 1958).

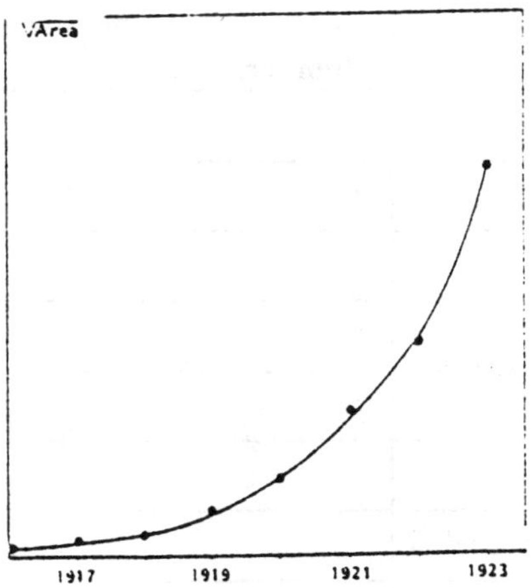

Figure 5b Second phase of the spread of the Japanese beetle into eastern North America (constructed after a map in Elton, 1958)

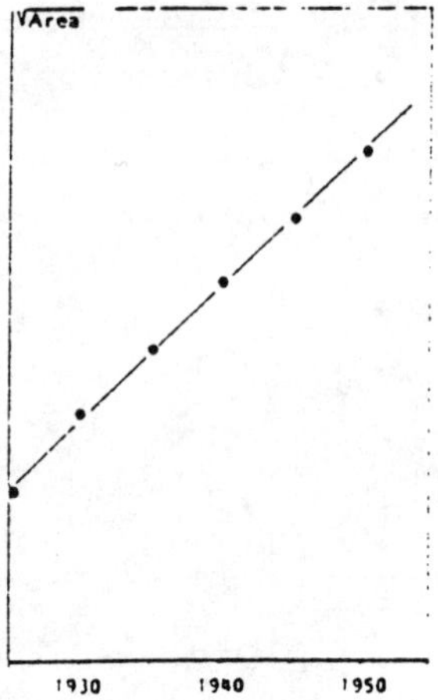

Simulation

We can also apply this model to slightly altered data for simulating the sensitivity of the results, depending on the kind of data alteration made (Hengeveld, 1992a, b; Marinissen & Van den Bosch, 1992). Earlier, I carried out three kinds of simulations, (1) altering the survival probability, (2) the amount of short-distance dispersal, and (3) the amount of long-distance dispersal. These simulation studies were not made for a technical sensitivity analysis of the model itself, however, but for testing the sensitivity of the biological system itself.

Table 1 Sensitivity of the expected velocity of spreading, C_{exp}, to changes in short-distance dispersal to the frequency of observed distances (x) in the Collared Dove.

N	**Number**	C_{exp} km^{-year}
68	35	57.49
69	36	57.09
70	37	56.69
72	38	56.30 (x)
73	39	55.92
74	40	55.55
79	45	53.79

Table 2 Sensitivity of the expected velocity of spreading, C_{exp}, to changes in the tail end of the observed (x) frequency distribution of distances covered by the Collared Dove.

Distance (km)						
N	**.575**	**.625**	**.675**	**.725**	**.750**	**C_{exp}**
69						43.8
70	1					48.0
71	1	1				51.9
71	2					51.6
72	2	1				55.6
72	1	2				56.3 (x)
72	1	1		1		57.6
72	1		2			57.6
73	1	2	1			60.6
73	1	1	1	1		61.9
74	1	1	1	1	1	66.6

Table 1 shows changes in expected invasion rate resulting from decreasing or increasing the number of short-distance dispersing Collared Doves relative to the actual observations, mainly affecting the number in the class of shortest distances covered. Decreasing this number in fact increases C_{exp}; increasing it decreases C_{exp}. The range of values of C_{exp} is relatively small.

Table 2 shows the sensitivity of C_{exp} relative to changes in the classes containing the jump dispersers. Decreasing the number of these dispersers decreases C_{exp}, and increasing it also increases the value of C_{exp}. Also, keeping their absolute number the same but redistributing them over the classes affects C_{exp} greatly. The range of C_{exp} is relatively large.

Comparing the results of Tables 1 and 2 shows that C_{exp} is more sensitive to long-distance dispersal than to dispersal over short distances. Moreover, this difference even affects the outcome of the species' sensitivity to changes in short-distance dispersal, inverting the sign of the outcome relative to the change made. Clearly, the effect of reducing the number of short-distance dispersers is indirect by giving them more weight to long-distance dispersal, thus increasing C_{exp}.

Interpretation

The changes made in the original data in this simulation study indicate the sensitivity of the invasion rate to naturally occurring changes in the landscape. The invasion rate, when conceived as a modified Bessel-function, consists of two components, the spreading capacity of the birds themselves and the trapping rate of the environment (Broadbent & Kendall, 1953), such as the number and configuration of suitable nesting trees. Changing the landscape implies changing C_{exp} through a change in trapping rate. Conversely, making the landscape more uniform means introducing a barrier to one species and an easy-to-penetrate environment to another. Also, as in the Muskrat, individuals can increase their own dispersal capacity by becoming more mobile under dry conditions. For both reasons, dispersal rate proves not to be a trait intrinsic to the species only, but to depend on environmental conditions as well. For a great deal, it is a variable, ecological species parameter. Only partly, it depends on traits as whether or not the animal can fly or swim, etc. how fast it can invade into a new region.

Yet, even the distinction between these abilities is not entirely independent of outward conditions; flying and walking, for example, proceed over a two-dimensional surface, whereas swimming can be channelled through linear ditches, rivers, or along a coast. This results in the first case in an exponential decrease of the number of propagules with distance and in the second case only in a linear decrease. With the same starting number, a species can cover only short distances in the first case and much longer distances in the second. The invasion rates therefore vary proportionally.

Similarly, the net rate of reproduction, R_o, varies as a function of ecological conditions, both according to changes in survival rate and to that of the individuals in the various age classes (Hengeveld, 1992a, b). This equally holds for some additional parameters taken up in a recent, more extended model (Van den Bosch et al., in press). These newly added parameters include, (1) the rate of settlement, (2) the amount of suitable habitat available in an area, (3) the survival rate of the settlers, (4) the length of the season for growth, reproduction, etc., and (5) the ratio of long- to short-distance dispersal.

Thus, the rate of spread of any species greatly depends on the variation in ecological factors. The degree of predictability, in turn, depends on the possibility and accuracy of estimating the relevant parameter values of a species under certain conditions, as well as on that of the variation in the factors affecting these values.

Limitations of the model

As any model, the presently applied one makes several assumptions that limit its application to those cases where these assumptions are met. If not, such as in the African bee, *Apis mellifera scutellata*, invading South and North America, respectively (Hengeveld, 1992a), the model fails. It can also fail to some extent in cases where jump dispersal dominates the process. The result of jump dispersal, namely, is the generation of a stochastic zone where the propagules settle by chance, surrounding the deterministically progressing invasion front by short-distance dispersal and population growth. Moreover, the settlement of jump dispersers is a spatially rare event, making it difficult to estimate, if at all possible. Therefore, we do not know what type of probability distribution applies to the dispersers at the tail end of the dispersal distribution, given that the generating process is structured. Because of these difficulties, the present model concerns the deterministic part of the process only.

The applicability of the model is also restricted by the assumption that environmental conditions are spatially uniform and temporally constant. In those cases where this assumption does not hold, the model cannot be applied. However, in some cases the scale of resolution can be adjusted such that variable conditions can be considered sufficiently constant. At the national scale of the former West Germany, for example, the Muskrat progressed at varying rates due to environmental variation, but at the broader, geographical scale of East and West Europe, the wave progressed steadily (cf. Hengeveld, 1989; Van den Bosch et al., 1992). It depends on the species, as well as on its spatial conditions if scale adjustment is feasible and realistic.

Finally, it is difficult, if possible, to apply the model to genetically engineered organisms as particularly ecologically sensitive parameter values have deliberately been altered, although in an unknown way. Moreover, other, unintended ecological side-effects can occur. It then depends on the degree to which organisms have been altered to which extent the parameter values of the parent organisms can still serve as a model for the newly engineered organisms; in principle, they cannot.

RISK ASSESSMENT IN BACILLUS THURINGIENSIS

Possessing a model working with independent, mechanistic parameters that has been tested with several species, does this allow us to assess risks possibly incurred by introducing genetic material from *B. thuringiensis* into the genome of a number of plant species? So far, we cannot, for several reasons.

First, we can look at what is known about the ecology of *B. thuringiensis* itself. Sadly, despite all our knowledge of its genetics, its physiology, and the way it interacts with its host (Aronson et al., 1986), virtually nothing is known about its ecology. The best-known parts of our knowledge concern the conditions under which it thrives in insects and the way it affects them (Aronson et al., 1986; Meadows, 1992). Its ecology as a free-living soil organism, however, as a possible part of its species ecology pertinent to its invasibility, is almost completely lacking. Conditions promoting population growth as one component of spatial spread, are unknown; dispersal as the other principal component seems to be limited, i.e. only covering a few metres (Meadows, 1992). Geographically, it is a real cosmopolitan species, found from the tundra to the steppe, which leaves no room for observations on local spread or on invasive behaviour. Ecologically, it is certainly euryoecious; it is only less common in beaches, caves, some desert locations, and rainforest (Meadows,

1992). There is even no correlation between the presence of insects and the number of isolates in which it was found. Epizootics occur only rarely (Burges, 1973); it is even not able to germinate in moist soils. The toxin itself degrades in circa three days due to cycles in temperature and humidity, proteolytic and microbial activity, photo-oxidation, and chemical interactions (Feitelson et al., 1990).

B. thuringiensis itself therefore seems not to incur any risk, neither to ecosystems, nor to various aspects of agriculture. This is particularly true since it does not affect vertebrates, whereas in the insect taxa (Lepidoptera, Diptera, and Coleoptera) and nematodes (Feitelson et al., 1992), its operation seems species specific as judged from direct observations and from inferences one can make as to its mode of action. Thus, after having been a successful biopesticide without any difficulties for about 30 years, its application as an entomopathogen seems to be safe and not to incur serious risk for its future use.

The risk of biopesticidal epizootics to occur may be greater after inserting the genes responsible for the formation of a certain type of the active insecticide, δ-endotoxin, into the genome of several crop species. Then, however, it depends on how and how much the parameter values of the host plant are altered, such that invasive behaviour becomes likely. Yet, which parameter values will be directly or indirectly altered, how much, and under what conditions will certainly not follow general rules from which expectancies can be derived in specific cases. It is reasonable to think of parameters determining the net rate of reproduction, R_o, but whether this will result from changes in survival of young or of old plants, from those in competitive power, or in fertility rate (cf. Tiedje et al., 1989) cannot be said without a detailed mechanistic knowledge of the species or variety concerned. Such knowledge, though, is completely lacking, particularly for plants (see Kooijman (1993) for first models developed for heterotrophes).

Assessing risks incurred by horizontal gene transfer through sexual reproduction is even more difficult, the risks of development of benign species into noxious ones being more unpredictable than the gene transfer itself. However, this kind of risk might be neglected in the context of that caused by genetic manipulation itself (Prins & Zadoks, 1993). Horizontal gene transfer is not known from transgenic plants carrying genetic material of *B. thuringiensis.*

Development of resistance against some strains of δ-endotoxin in certain insects (e.g. the moth *Plodia interpunctella*) (Gould, 1988; McGaughey, 1990; McGaughey & Whalon, 1992) is a relatively recent complication in risk assessment of genes of *B. thuringiensis* inserted in plants. Strains of *P. interpunctella* resistant to spores and crystals of delta-endotoxin were still susceptible to exotoxin (McGaughey, 1990). Then, the risk also depends on the spatial dynamics of the insects, as well as on spatial aspects of their genetic resistance development. It should be realized, however, that population-genetic models have not yet been worked out within a spatial context (cf. May & Dobson (1986) and May (1993) for some first thoughts in this direction). One option at the physiological level is to develop mixtures of protoxins that both broaden the specificity and enhance the activity on the pest species concerned (Aronson & Wu, 1990). As a possible countermeasure against the resistance development, one can think of reserving refugia of untreated plants in which resistance does not develop and from which gene flow occurs to dilute the pool of resistant insects, thus retarding the process (May, 1993). Again, it depends on invasion parameters, but now on those of the insect, at what density and configuration such refugia have to be laid out to be effective, what size they should optimally have, and how those spatial parameters depend on the field size of the insect population actually developing resistance. Apart from appropriate models still not being developed, the practical sides of parameter estimation should not be underestimated. Spatial sampling is extremely laborious, as is the subsequent species identification and data processing, let alone spatial monitoring on a regular basis.

Bacillus thuringiensis as an experimental model

The more complicated, mathematical models needed in this context cannot be developed without testing the various assumptions along the way. Part of these assumptions can be tested using information from the literature or obtained from the laboratory. Another part, however, should be drawn from the field. Moreover, the applicability and sensitivity of the overall model should eventually be tested by the exclusive use of field data.

However, working with species or varieties the ecological parameter values of which have deliberately been altered (such as unaffected fruits or leaves of tomato plants containing endotoxin in their genome, which not only may please potential consumers, but which can also improve the plant's health and hence its ecological performance), any field testing is risky and should, as a consequence, be minimized. Ideally, model development should be accompanied by field experiments on organisms that are regarded safe at the level of the individual organism, rather than on those still bearing severe potential risk. Also, effects of countermeasures to be taken and derived from the model should, ideally, be estimated using organisms that are safe at the individual level first.

B. thuringiensis, by now having a safety record of circa 30 years of field application under widely different conditions, seems to meet this safety requirement very well. Moreover, it is known by many farmers who could collaborate by doing field observations, for example, by collecting potentially resistant insects. They should also agree and collaborate using chemical pesticides in genetic refugia from where non-resistant insects disperse into areas with resistance development. As *B. thuringiensis* is still not universally applied, effects should be studied of varying the number, configuration, size, mutual distance of such genetic refugia, or of the combination in which chemopesticides and biopesticides are applied within a certain region. Also, alterations in application of both types of pesticide may be necessary - at least during an experimental phase - and should be feasible for *B. thuringiensis* under present conditions of use.

Thus, rather than accepting *B. thuringiensis* as sufficiently safe for broad application and doing the necessary experiments on species or varieties with less-known properties and a potentially higher risk, *B. thuringiensis* should be considered to be used as experimental tool for large-scale spatial processes to be modelled.

DISCUSSION

Ultimately, after hazard identification, all risk assessment concerns predictability. Yet, it seems important to be clear about what we actually mean how and how closely to predict an invasion to occur. The difficulty is that there are several levels of prediction, the one closer than the other. It then depends on the problem at hand which level one wants to choose. However, the various levels are not independent; they are nested. Thus, for a precise prediction, one should also be able to predict at one or more of the coarser levels.

The coarsest level is to predict whether or not a transgenic plant species can spread into a new region anyway, depending on local climate, topography, or soil conditions, etc. (Forcella & Wood, 1984). This prediction depends on the species' physiological response to abiotic components in its environment. Usually, one estimates the species' preferences in its region of origin and projects its location in climate space into the new geographical space (see Hengeveld (1990) for the use of these concepts). When climate or local conditions vary, the invasion speed will vary accordingly, as indicated above for the Muskrat in West Germany. The assumption here is that the species' response remains the same, which is intentionally violated in transgenic organisms, making prediction of field behaviour difficult.

The second level of prediction concerns a more intrinsic component determining the velocity rate of invading. This component concerns the species' morphology and behaviour. It explains the scatter along the 45° line in Figure 2. Of course, prediction at this level is not restricted to species properties alone, but it also includes environmental components, such as landscape structure. Thus, the rapid spread of the Cattle Egret, *Bubulcus ibis*, in Figure 2, is partly explained by the species' dispersal capacity and partly by the continuity of cattle farming in North America. The same holds for various epidemic diseases in agricultural monocultures.

The next level of prediction concerns the species' ecological and behavioural properties related to its life history and dispersal behaviour. This level concerns predicting the rate of spread after a known species has been introduced into a new region or continent, such as the recent introduction of the Collared Dove into Florida (Hengeveld, in press). This way of predicting therefore depends on components determining its spatial demography. As shown above in the simulations for the Collared Dove, again prediction does not solely depend on properties intrinsic to the species but on an interaction between these properties and characteristics of the environment. It predicts the tightness of the scatter of points around the 45° line in Figure 2. This tightness can be enhanced by including more and more parameters into the model (see above), although parameter-sparse models remain to be preferred. If most points lie above or below the 45° line, one may have excluded a single factor or process. Thus, most points in Figure 2 tend to lie above the line, because long-distance dispersal resulting in the stochastic zone ahead of the closed front was not accounted for in the model (Hengeveld, in prep.).

The fourth level of prediction takes the dependence on other species into account. Thus, the spread of transgenic plant species, including genetic material of *B. thuringiensis* in their genome, depends in part on genetic resistance development in a herbivorous insect. This development, in turn, depends on the predictability of the population genetics of this pest insect, as well as on that of its spatial spread at the various levels of prediction. As indicated, this is still *terra incognita*, models of gene flow still including the dispersal component of spatial spread only and lacking the demographic one. For another part, spread of a transgenic plant depends on its competitive power relative to that of other species. This power seems to be limited as judged from one single field experiment so far (Crawley et al., 1993; Rees et al., 1991). Spread including other species is therefore not predictable at this level.

These various ways of prediction still do not include testing. There are two ways of testing, descriptive, statistical testing and purpose testing. In descriptive testing, one can, as a first measure, estimate the closeness of prediction C_{obs} from C_{exp} by taking their ratio and expressing the result as the percentage of variance thus explained. Statistically, one would like to calculate confidence limits around the 45° line in Figure 2, for which, however, the number of points is still too small. Yet, it contains all vertebrate species known to me in the literature with the information required. Purpose testability considers some aim one has in mind, seeking to test the fit between this aim and the results obtained. Thus, in the context of genetically engineered organisms, one may want to know whether or not a species approaches a predefined risk according to its demography, or to its spreading rate or potential. One starts, therefore, with a model of some overall process the sensitivity of which is known regarding the parameter to be tested. The question then becomes how to assess the risk of a particular species behaviour within the framework of this overall process. One needs to know both the mechanistic prediction of the particular spreading process, as well as that of the overall process and their interaction. Predictability from the viewpoint of purpose testing is not possible as yet in the field of risk of genetic engineering at the population level.

Thus, because these prediction levels are nested, the criteria for modelling, observation, and treatment are given by the process at hand; they do not follow without structure from intuitively chosen ecological concepts as in Crawley (1990) or Tiedje et al. (1990). This choice also depends, however, on the economic cost involved. The closer the prediction, the higher the cost. At the present stage, prediction at the population level as described in this paper, will be very expensive, not only because the parameter values required need to be estimated - as will always be necessary - but particularly because most new techniques of estimation - involving experimentation, spatial observation and data processing, and modelling - have to be developed. As soon as this has been achieved, spatial monitoring needs to be done for each process related to a certain level of prediction of the transgenic organism concerned, adding to structural cost.

This approach of stepwise refining of risk assessment as viewed against the background of economic cost can also be broadened to the cost of the risk of genetic impoverishment as a potential consequence of genetic engineering. For practical reasons, namely, only one or a few of the many existing crop varieties will be manipulated, the new variety subsequently economically outcompeting all others over broad continental expanses, or even world-wide. The resulting, genetically homogeneous crop will be exceptionally vulnerable to the pest development, as well as to the development of insect resistance against many pesticides. Moreover, apart from this greater vulnerability, the few remaining races will lack the amount of variability necessary for coping effectively with other environmental risks they run. It seems to me that this sort of risk incurred by genetic engineering may deserve even more attention than any other, more immediate risk from the ecological improvement itself.

CONCLUSIONS

This paper analyzed the invasion process, describing it at a reasonably close level in mathematical terms of a recent model on the process mechanism. It appears from a simulation study that the invasiveness of a species is not a trait intrinsic to the species but that it reflects the species' success to match favourable environmental conditions. We should know, therefore, both the species' ecological properties and the environmental conditions to be able to assess this species' risk of invasion into some new region. As transgenic organisms are intended to show novel ecological responses, risk assessment at the population level will be extremely difficult, if possible. Additional difficulties arise from the stochastic nature of part of the invasion process, i.e. of the impact of jump dispersal and, in the case of *B. thuringiensis*, from ecological properties, dynamics, and resistance development in the insects affecting transgenic crop plants. No models are available to allow for spatial aspects of resistance development from which to derive countermeasures to be taken. As *B. thuringiensis* has proved to be safe at the level of the individual, it could be used as a model for field experiments on resistance development.

REFERENCES

Aronson, A.I., W. Beckman & P. Dunn (1986): *Bacillus thuringiensis* and related insect pathogens. Microbiol. Rev., 50: 1-24.

Aronson, A.I. & D. Wu (1990): Specificity *in vivo* and *in vitro* of *Bacillus thuringiensis* delta-endotoxins active on Lepidoptera. In: R.R. Baker & P.E. Dunn, eds. New directions in biological control, p. 547-560. Liss, New York.

Barrows, W.B. (1889): The English sparrow in North America, especially in its relation to agriculture. USDA, Div. Econ. Ornithol. Mammal. Bull., 1: 1-405.

Broadbent, S.R. & D.G. Kendall (1953): The random walk of *Trichostrongylus retortaeformis*. Biometrics, 9: 460-466.

Burges, H.D. (1973): Epizootic diseases of insects. In: L.A. Bulla, ed. Regulation of insect populations by microorganisms, p. 31-49. Ann. New York Acad. Sci., 217. New York.

Cliff, A.D., P. Haggett & G.R. Versey (1981): Spatial diffusion. Cambridge University Press, Cambridge.

Crawley, M.J. (1990): The ecology of genetically engineered organisms: assessing the environmental risks. In: H.A. Mooney & G. Bernardi, eds. Introduction of genetically modified organisms into the environment, p. 133-150. Wiley, Chichester.

Crawley, M.J., R.S. Halls, M. Rees, D. Kohn & J. Buxton (1993): Ecology of transgenic oilseed rape in natural habitats. Nature, 363: 620-623.

Elton, C.S. (1958): The ecology of invasions by animals and plants. Methuen, London.

Feitelson, J.S., J. Payne & L. Kim (1992): *Bacillus thuringiensis*: insects and beyond. Bio/Technology., 19: 271-275.

Feitelson, J.S., T.C. Quick & F. Gaertner (1990): Alternative hosts for *Bacillus thuringiensis* delta-endotoxin genes. In: R.R. Baker & P.E. Dunn, eds. New directions in biological control, p. 561-572. Liss, New York.

Fisher, J. (1953): The collared turtle dove in Europe. Brit. Birds, 56: 153-181.

Fisher, R.A. (1937): The wave of advance of advantageous genes. Ann. Eugen., 7: 355-369.

Forcella, F. & J.T. Wood (1984): Colonization potentials of alien weeds are related to their "native" distributions: implications for plant quarantine. J. Austr. Inst. Agric. Sci., 50: 35-41.

Gould, F. (1988): Genetic engineering, integrated pest management and the evolution of pests. TREE, 3: 515-518.

Haggett, P., A.D. Cliff & A. Frey (1977): Locational analysis in human geography. 2 Vols. Arnold, London.

Hengeveld, R. (1988): Mechanisms of biological invasions. J. Biogeogr., 15: 819-828.

Hengeveld, R. (1989): Dynamics of biological invasions. Chapman & Hall, London.

Hengeveld, R. (1990): Dynamic biogeography. Cambridge University Press, Cambridge.

Hengeveld, R. (1992a): Potential and limitations of predicting invasion rates. Florida Entomol., 75: 60-72.

Hengeveld, R. (1992b): Cause and effect in natural invasions. In: J. Weverling & P. Schenkelaars, eds. Ecological effects of genetically modified organisms, p. 29-43. The Netherlands Ecological Society, Arnhem.

Hengeveld, R. (1994): What to do about the North American invasion by the Collared Dove? J. Field Ornithol. (in press).

Hengeveld, R., & F. Van den Bosch (1991): The expansion velocity of the Collared Dove (*Streptopelia decaocto*) population in Europe. Ardea, 79: 67-72.

Kirk, T.W. (1890): Note on the breeding habits of the European sparrow (*Passer domesticus*) in New Zealand. Trans. Proc. New Zealand Inst. Zool., 23: 108-110.

Kooijman, S.A.L.M. (1993): The dynamic theory of energy budgets. Cambridge University Press, Cambridge.

Marinissen, J.C.Y. & F. Van den Bosch (1992): Colonization of new habitats by earthworms. Oecologia, 91: 371-376.

May, R.M. (1993): Resisting resistance. Nature, 361: 593-594.

May, R.M. & A.P. Dobson (1986): Population dynamics and the rate of evolution of pesticide resistance. In: E.H. Glass *et al.* Pesticide resistance: strategies and tactics for management, p. 170-193. National Academy Press, Washington.

McGaughey, W.H. (1990): Insect resistance to *Bacillus thuringiensis* δ-endotoxin. In: R.R. Baker & P.E. Dunn, eds. New directions in biological control, p. 583-598. Liss, New York.

McGaughey, W.H. & M.E. Whalon (1992): Managing insect resistance to *Bacillus thuringiensis* toxins. Science, 258: 1451-1455.

Meadows, M.P. (1992): Environmental release of *Bacillus thuringiensis.* In: J.C. Fry & M.J. Day, eds. Release of genetically engineered micro-organisms, p. 120-136. Cambridge University Press, Cambridge.

Nowak, E. (1965): Die Türkentaube. Ziemsen, Wittenberg-Lutherstadt.

Nowak, E. (1989): Ausbreitung der Türkentaube (*Streptopelia decaocto*) in der UdSSR: Umfrage 1988. J. Ornithol., 130: 513-527.

Pielou, E.C. (1979): Biogeography. Wiley, New York.

Prins, T.W. & J.C. Zadoks (1993): Horizontal gene transfer in plants, a literature review. NRLO Report, 93/15, Wageningen.

Rees, M., D. Kohn, R. Hails, M. Crawley & S. Malcolm (!991): An ecological perspective to risk assessment. In: D.R. MacKenzie & S.Z. Henry, eds. Biological monitoring of genetically engineered plants and microbes. Proc. Kiawah Island Conf., November 27-30, 1990, p. 9-24.

Schröpfer, R. & C. Engstfeld (1983): Die Ausbreitung des Bisams (*Ondatra zibethicus* Linné, 1766, Rodentia, Arvicolidae) in der Bundesrepublik Deutschland. Z. angew. Zool., 70: 13-37.

Skellam, J.G. (1951): Random dispersal in theoretical populations. Biometrika, 38: 196-218.

Stresemann, E. & E. Nowak (1958): Die Ausbreitung der Türkentaube in Asien und Europa. J. Ornithol., 99: 243-296.

Tiedje, J.M. *et al.* (1989): The planned introduction of genetically engineered organisms: ecological considerations and recommendations. Ecol., 70: 298-315.

Van den Bosch, F., R. Hengeveld & J.A.J. Metz (1992): Analysing the velocity of animal range expansion. J. Biogeogr., 19: 135-150.

Van den Bosch, F., J.A.J. Metz & O. Diekmann (1990): The velocity of spatial population expansion. J. Math. Biol., 28: 529-565.

Van den Bosch, F., J.C. Zadoks & J.A.J. Metz (in press): Continental expansion of plant disease: a survey of some recent results.

Zawolek, M.W. & J.C. Zadoks (1992): Studies on focus development: an optimum for the dual dispersal of plant pathogens. Phytopathol., 82: 1288-1297.

PLENARY DISCUSSION

Question:
This presentation was a perfect example of what I like about presentations by ecologists. First, they scare you to death by saying that we do not know enough, that there too many unknowns to predict anything and then, finally, they end by saying there is no problem at all. It is a kind of pattern I have experienced with discussions with several ecologists: We don't know enough but don't worry! One sentence, and you have repeated it twice, so I cannot have misunderstood it, is: Invasiveness is not an intrinsic trait. Given the fact that in current risk assessments we are all looking for selective advantage, is the conclusion of what you said that in fact we should not look at that anymore? Because you said invasiveness simply depends on the environment; there are no intrinsic characteristics of the organism contributing to its invasiveness, whereas those, who are looking at risk assessments on a daily basis, look, among other things, at selective advantage because that might contribute to the organism's invasiveness. Could you elaborate a little more on this point?

Response:
The net reproductive rate needs to be just over 1, when it would be constant, and that can be achieved in very different ways. This is actually the thing we are talking about. Whether or not you are producing an enormous number of seeds or eggs or whatever, if there is an enormous mortality counterbalancing it at that stage, then you still can have a negative result; a net reproductive rate just below zero. What is really important is to get just over it. The example I gave of the collard dove that lays only two eggs per clutch but repeatedly over the year up to six times and therefore it can do it. If it also knows to do it over many years, then it has a net reproductive rate of 1.33. This is still low, at least in my view; net reproductive rate explains its enormously fast rate of expansion across Europe. So, it is not the number of propagules or seeds or eggs but it is the balancing between the mortality and the reproductive rate.

Question:
I am not quite satisfied. My question goes one step further. We are dealing with all kinds of cases -let us say biotic stress resistance, drought tolerance or other characteristics. As far I understand you, those characteristics could be forgotten in the whole risk assessment, whereas the discussion is that abiotic and biotic stress resistances may contribute to invasiveness. And what you are saying is: Forget about that and let us talk about net reproduction rates and that is it.

Response:
That is actually the most important thing we are talking about and this net reproductive rate is influenced by ecological factors that are very species-specific. It depends, for example, at what stage mortality is highest in the first stage of life or not; that makes a difference. I gave you this life table and there you saw the survivor rate. Well, if that is constructed differently, you will get a different net reproductive rate. If the fertility distribution over the age classes differs, you will get

another one. And both are very much dependent on ecological conditions. They also vary from one species to another. So, we just cannot generalize about this.

Question:
I am reminded of early discussions in the OECD coming up with GDP (Good Development Practice), where we did not try to categorize an organism but rather the trait, the organism, and the environment into which it was to be introduced. I think that is just what you are saying: It is not just the trait, it is dependent of the ecology and the location where the test is conducted.

Response:
Yes indeed. In relation to the predictability, we now can reconstruct, under certain conditions, these invasions but we cannot predict under different conditions the invasiveness of the same species. We just do not know how it will respond under different conditions.

Question:
We do not want to get too pessimistic about this. If you look at your graph, you can draw from it some substantial generalizations. For example, nearly all of the birds that you talked about were generalists in terms of diet; they have very broad diets, a lot of seed and insect feeding birds there, a lot of them directly associated with man like the house sparrow, or indirectly associated like the starling and the collared dove. And those kinds of environments come out very strongly when you look at invasible kinds of communities. On the other hand, I do not want to destroy the one generalization you did have which was that the rates of radial spread were species-specific for these birds. That is completely true. But we have just completed an unpublished study on the twenty alien micro Lepidoptera that have invaded Great Britain. All those species show exactly the same rate of radial spread, even though they are different-sized by a factor 4 or 5. So, different species having different rates of radial spread is not a robust generalization across different taxa, I am afraid.

Response:
I fully agree with you on the first remark about generalized nature of the species; that is true. They are actually the best known. But you also know that there are now many and many species expanding over Europe. However, we do not have the basic data to make similar calculations. Since the last war, 150 of the 500 bird species in Europe have expanded and not all of these are generalized species. In relation to the predictability of the expansion of generalized species: What is a generalized species? *Eichornia*, for example, has expanded enormously across the tropics but a very related species is still sitting, almost endemically, in the Amazon region. So far nobody has come up with an answer to why these very similar species do not behave in the same way. So, I do not know what a very generalized species is. I cannot predict that and cannot explain it afterwards either. Can you?

Question:
Until now the discussion has centred around invasiveness by comparing the same or different species in the same or different environments. But when we are talking

about the relevance of this to GMOs, we should, in fact, be looking more at varieties, that is to say organisms that are genetically almost identical, only with a few differences. My question therefore to you and others here is: How many studies of this kind have been done, comparing different varieties rather than different species?

Response:
The set of species I have shown you in this graph were vertebrates. This is all we have in the total ecological and biogeographical literature. I have not come across any other example. But perhaps my colleague knows more about fungi.

Intervention:
Yes, we have quite a bit of information on the invasiveness of parasitic fungi at the varietal level. And the pattern is always the same: It is not very different from the parasitic species itself. So, within the species the varieties behave very similarly.

CONCLUDING REMARKS ON INSECT RESISTANT PLANTS

Prof. dr. S.A. Tolin (rapporteur), Department of Plant Pathology, Physiology & Weed Science, Virginia Polytechnic Institute & State University, United States of America

SUMMARY

It was concluded that the development and use in agriculture of annual crop plants expressing low levels of the Bt-toxin to confer resistance to susceptible insects has no potential for causing long-term ecological impacts. This is because no impact is expected on any keystone species, and no relatives of crop plants to which the gene might transfer are known to be affected by herbivores sensitive to the toxin. Failing to adopt Bt-toxin resistance technology as a part of pest management strategies in crop plants could negate the opportunity to decrease chemical insecticide use. The primary issue is that approaches and strategies for applying the technology so as to retain efficacy, in the face of constantly evolving insect populations, need to be devised.

Discussions of the impact of plants genetically modified to resist insects was limited to cases in which the δ-endotoxin from *Bacillus thuringiensis*, and other currently considered modifications related to this case, had been inserted into the genome of an annual crop plant. For these cases, the discussants surmised that "long-term" was a period of time greater than ten years, and that an ecological impact was an event that had a major impact on a keystone species. It was concluded that the Bt toxin constructs in the crop plants that have been tested to date are not expected to exhibit any long-term ecological impact.

Crop plants are unusual in that they are uniform and often susceptible to insects, and large populations of plants and plant products allow specific insect species to reach large populations and become pests to those plants. However, even without addition of the Bt gene, most non-crop plants are toxic to most all insects. Crop plants tested to date, with few exceptions, are not expected to transfer the inserted gene to wild plants that are relatives. Even if gene transfer were to occur, no relatives that might be expected to receive the gene are currently affected by herbivores that would respond to the Bt gene.

Application of Bt-resistance technology is likely to occur worldwide in several crops. The driving force behind the development and use of the technology is the desire to change agricultural pest control practices to reduce the use of chemical insecticides having known ecological impacts. Therefore, a concern for Bt-resistance technology is continued efficacy of the
product. As when chemical insecticides are used, insects have in isolated cases developed resistance to bacterial form of *Bacillus thuringiensis* when used widely as a biological control agent.

Biotechnology companies that have developed and will be the suppliers of insect resistant plants have an interest in keeping a durable, useful product on the market.

The need for a strategy to temper the build-up of resistant insect populations is well recognized. Discussions are currently ongoing among the users of this technology to devise the best integrated approaches to manage pest populations through the use of resistant plants and insecticides. Research projects are underway worldwide on this phenomenon. The goal is to achieve pest management that is durable, ecologically-sound, yet economically profitable for the company and the farmer.

One of the factors permitting the conclusion that there is expected to be no long-term ecological impact of Bt-technology is that the use of insect resistant plants in agricultural systems will be continuously monitored. It was judged appropriate that this could be done by the farmers, the users of the insect-resistant plants, and by the developers and suppliers of the plants, for reasons mentioned above. Of the currently-used risk assessment factors, survival, persistence and dissemination were judged only moderately applicable to monitoring the long-term impact of insect resistant plants. The most important component was thought to be invasiveness of the plant, but it was recognized that invasiveness may not necessarily be negative in ecological terms.

A second issue, which is more of an efficacy issue, is the impact on insect populations, most probably build-up of resistance. Evolutionary changes in insect populations in response to pest control measures is occurring now in agricultural systems. The potential for change induced by plant-delivered Bt was not judged to be severe a concern, incrementally, that it should limit the introduction of the Bt-technology into agricultural systems. However, the alternative of not using the Bt-technology has a high potential ecological impact since, presumably, chemical insecticides would continued to be used.

Since the conclusion was that no long-term ecological impact was expected, no recommendations could be directed toward alleviating the impact. It was recognized that some species in an ecosystem may be affected, but other species may be benefited. Applying the term "harm" to such an affect is not consistent with ecological concepts. Indeed, it was stated that harm is not a concept that ecologists place a value on. Rather, harm is a socio-economic issue.

Finally, using the above arguments, judicious application of Bt-resistant plants should take into consideration the use of molecular approaches to design plants whose resistance is less likely to be overcome by evolving insect populations. One of the key issues is that only a very low expression of the Bt gene appears to be needed for efficacy. A second is that specificity for target insects is quite high. The use of multiple toxin genes and appropriate promoters is a possible strategy. Combining the expression of Bt with, for example, a promoter stimulated by insect feeding, such as the proteinase inhibitor gene's promoter, is being examined by some researchers as a means to express the gene only as needed. It was concluded that combination of the Bt toxin gene with other genes would cause no change in the assessment of the long-term ecological impact of insect resistant plants.

TOPIC 2

VIRUS RESISTANT PLANTS

AN INTRODUCTION TO DISCUSSION OF THE POTENTIAL LONG TERM EFFECTS OF TRANSGENIC PLANTS EXPRESSING VIRAL GENES

Dr. M. Tepfer (rapporteur), Laboratoire de Biologie Cellulaire, Institut National de Recherche Agronomique, France

Plant viral diseases cause serious crop losses world-wide. For this reason, an important feature of plant breeding has been the introduction of natural virus resistance genes into cultivated genotypes, often from closely related wild plant species. However, in many cases, usable natural resistance genes are not known, in which case control of spread of plant viruses is often based on less efficient means, such as the use of pesticides to reduce populations of the most common vectors of viral disease, insects and soil fungi.

In 1986, R. Beachy and co-workers showed that expression of a viral coat protein gene in transgenic plants confers resistance to the virus from which the gene was derived. Since then, coat-protein-mediated resistance in plants has been obtained for at least twenty plant viruses. More recently, it has been shown that expression of other parts of a viral genome in transgenic plants can also confer resistance. This is the case, for instance, for certain genes encoding either native or modified forms of a viral replicase subunit, as well as for ones encoding a viral protein necessary for cell to cell spread of the virus. Expression of non-translated viral RNA, including segments of the viral genome that are sites of initiation of replication, can also interfere with virus infection. Of known competitor RNAs, the most extensively studied have been the satellite RNAs associated with certain strains of cucumber mosaic virus (CMV); several laboratories have shown that plants expressing satellite RNA genes are protected against CMV.

Though these pathogen-derived resistance genes shown great promise, there is also concern that expression of viral genes in plants may have a potential for long-term negative impacts. This was first proposed for plants expressing CMV satellite RNA genes, since certain satellite RNAs can cause symptom worsening rather than protection, and it has been shown that a point mutation in a beneficial satellite RNA can render it deleterious to certain hosts in the presence of CMV. It is less obvious that other types of viral genes could have a negative impact, but two mechanisms by which they could affect the composition of viral populations have been proposed. It has been shown that heterologous encapsidation can occur in plants expressing a coat protein gene. In this case the genome of an infecting virus is encapsidated in the coat protein (of a more or less different virus) synthesized by the plant. Since the coat protein is the site of recognition of virus particles by the natural vectors of virus transmission, the hetero-encapsidated virus will have acquired the potential for transmission of the coat protein synthesized by the plant. It should be noted that in this case, the viral genome is not modified and the new characteristics are lost when the virus infects a host plant and synthesizes the coat protein encoded by its own genome. A more general potential mechanism for plants expressing any type of viral gene to have an effect on the composition of virus populations is the possibility for recombination between the transcript expressed by the plant and the infecting virus.

In contrast to hetero-encapsidation, this would lead to durable modification of the viral genome and would be a mechanism for generation of new types of virus.

Three different mechanisms are described above by which plants expressing viral genes could have an impact on virus populations: mutational drift in transcripts of CMV satellite RNA genes, hetero-encapsidation in plants expressing coat protein genes, and recombination between transcripts of all types of viral genes introduced into the plant genome and an infecting virus. The first two mechanisms have been shown to be possible under controlled conditions in the laboratory. In the third case, there is evidence for recombination between cellular and viral RNAs having occurred naturally in the past, and in fact, recombination between viral genomes has been demonstrated in the laboratory, and has been proposed as a general mechanism in RNA virus evolution.

Though these potential mechanisms have been identified, we are at this time unable to evaluate clearly the real impact of plants expressing viral genes on natural virus populations and the epidemiology of virus disease. One difficulty arises from the limits of our knowledge of the frequency of mutational drift, of hetero-encapsidation and of RNA recombination. An important point of reference is that infection of plants by more than one virus is a relatively frequent event. In this case, one might expect that there would have been ample opportunity for hetero-encapsidation or recombination to occur. Would large-scale cultivation of plants expressing viral genes lead to effects in addition to those caused by natural co-infection? To some extent, the answer will depend on exactly which gene is expressed in which plant. For instance, in cases where viral sequences are introduced into plants that are not natural hosts of the virus, this might allow more novel events of hetero-encapsidation or recombination to occur.

In any case, the above considerations are all of relatively short term events. A far more difficult question is to what extent hetero-encapsidated viruses or new forms created by recombination or mutation could over a longer time scale affect the epidemiology of plant viral diseases. Studies of the fate of new molecules within a viral population, and epidemiological studies of non-modified viruses should provide useful clues.

PLANT VIRAL GENES INTRODUCED INTO THE HOST GENOME TO OBTAIN VIRUS RESISTANCE: THE PAST, THE PRESENT AND THE FUTURE

Dr. E. Balázs, Agricultural Biotechnology Centre, Hungary

SUMMARY

In the last decade, since the first successful demonstration of virus disease resistance in transgenic plants, a new research area has started to prosper. Starting with coat protein mediated resistance several alternative ways have been discovered, including the use of satellite RNAs, the antisense approach, catalytic RNAs, replicase mediated protection, and the use of engineered defective interfering RNAs. All these strategies have proved to be useful in small scale laboratory experiments or in greenhouse conditions. Field experiments have been performed more and more widely all over the world and have tended to become large scale experiments. This has led to an analysis of the potential risk associated with transgenic plants bearing viral sequences. The three major possible risks are considered to be, transencapsidation, synergistic interactions and recombinations. In the light of related publications and experiments, this review summarizes the connection between different technologies using viral sequences on engineered virus resistance and the potential risks associated with them.

INTRODUCTION

Accurate figures for crop losses due to viruses are not available, but data from epidemiological studies indicate the importance of these pathogens in agriculture and the environment. Tristeza, as the most important virus disease of citrus has killed more than 50 million trees all over the world (Bar Joseph et al., 1981). It is difficult to obtain reliable data on losses, but the extent of direct losses with different crops in different seasons and in various regions can be formulated. However, no curative technologies have been developed for plant protection against viruses. Only preventive measures are used in practice. The discovery of cross protection (McKinney 1929) has led to experiments which have confirmed that deliberate early infection would be worthwhile for a farmer who has always suffered quality loss. In this case, infection of a plant with a strain of virus causing only mild disease symptoms may protect it from severe strains. There are several successful examples of the use of cross protection for cherry tomatoes in the UK, for greenhouse grown tomatoes in The Netherlands or for citrus worldwide. In spite of these very spectacular results, it is not to be recommended as a general practice because of epidemiological, environmental considerations. The three major arguments are:

(i) when an unrelated virus is introduced into the plants, a serious disease may develop due to the mixed infection;

(ii) the protecting virus strain can mutate to a more severe type on other plant species;

(iii) the protected crop may act as a reservoir of a virus from which other species or varieties can become infected.

The tobacco mosaic virus mild strain can protect tomatoes from disease symptom development, while the same strain is lethal to pepper, which suggests the limitations of this technology.

At present, genetic engineering has made it possible to envisage the development of protection against these pathogens. Our knowledge of the genome organization of plant viruses has evolved rapidly in the last decade and has helped to build strategies for protection. Since the first successful experiment on coat protein mediated cross protection (Powell et al., 1986), more than a hundred host-virus relationships have been involved in this technology (Table 1). In the case of this first successful experiment, the transgenic plants expressed the coat protein gene of the introduced virus and conferred resistance against the same virus and against its strains.

Table 1 Selected references on coat protein mediated cross protection

HOST	VIRUS	REFERENCES
tobacco	AIMV	Tumer et al. 1987
tomato	AIMV	Tumer et al. 1987
alfalfa	AIMV	Halk et al. 1989
tobacco	CMV	Cuozzo et al. 1988
tobacco	TSV	van Dun et al. 1988
potato	PVS	Mackenzie et al. 1990
tobacco	PVS	Mackenzie et al. 1990
beet	BNYVV	Kallerhof et al. 1991
potato	PLRV	Kawchuk et al. 1990
tobacco	PLRV	Kawchuk et al. 1990
tobacco	TSWV	Gielen et al. 1991
tobacco	PVX	Hanenway et al. 1988
potato	PVX	Fehér et al. 1992
tobacco	PVY	Kollar et al. 1993
potato	PVY	Lawson et al. 1989
tobacco	SMV	Stark et al. 1989
tobacco	TRV	van Dun et al. 1988
tobacco	TMV	Powell et al. 1986
tomato	ToMV	Sanders et al. 1991
corn	MDMV	Murray et al. 1993
corn	MCMV	Murry et al. 1993

Today, almost all of the primary structures of the important plant viruses are known, leading to the use of the coat protein gene constructs. In most cases, this is relatively easy because the coat protein genes are encoded by subgenomic RNAs, while in certain viruses it is more difficult. This is the case with potyviruses, where the coat protein is part of a larger polyprotein.

Using site specific mutagenesis, a precise sequence of the coat protein gene can be obtained. These sequences have to be cloned into a transformation vector with an appropriate transcriptional promoter and termination signals. The almost exclusively used cauliflower mosaic virus 35S promoter has turned out to be the strongest one. Early experiments with the tobacco mosaic virus coat protein gene suggest that the amount of expressed coat protein gene product is very important to achieve sufficient resistance against challenge infection (Bewan et al., 1985, Powell et al., 1986). However, recent results with potato virus Y coat protein constructs, where truncated or antisense coat protein genes have been introduced into plants, cross protection or in certain cases immunity has been observed (van der Wilk et al., 1991, Farinelli and Malnoe 1993). Efficient coat protein mediated cross protection was induced by the integrated potato virus Y coat protein gene in several tobacco lines, while from the same experiments nonresistant tobacco lines were found contrary to the expression of the integrated coat protein gene (Kollár et al., 1993). This suggests that the observed resistance is independent of the level of coat protein accumulation in that transgenic tobacco. These data also suggest that there are different mechanisms for the coat protein mediated cross protection in different virus-host combinations. The resistance is manifested in a local lesion host of the virus with fewer infection sites on inoculated leaves, while in systemic hosts systemic spread of the virus is reduced by delaying the disease development. One of the most important results is the low accumulation of the challenge virus in transgenic plants. This fact has an important epidemiological effect by reducing the virus titer in the reservoir plants.

Besides the laboratory experiments, a high level of virus resistance has been detected in field conditions for potato, tomato and other important crops. Field experiments have already been performed in more than twenty countries to evaluate the efficiency of this control measure. ORSTOM has launched an international cassava-trans project on cassava genetic engineering to obtain virus resistance. This project proposes to apply modern biotechnology to control cassava viral diseases, concentrating on the two major virus diseases of cassava, namely the African cassava mosaic virus and the cassava common mosaic virus. A similar global project has been initiated and sponsored by the Rockefeller Foundation to produce genetically engineered virus resistant rice.

The use of this technology for developing virus resistant crops is well documented. In the near future, more examples with different viruses and hosts will be reported and even some transgenic crops become commercially available after meeting the regulatory requirements. In the laboratory, extensive research will be performed to study the mechanisms of this coat protein mediated resistance.

The term coat protein mediated resistance became part of the concept of the Sanford and Johnston (1985) classification of pathogen derived resistance. Viruses have become the number one target of this concept, based on their small genome size and their relatively simple organizations.

A surprisingly high level of virus resistance was found when the putative replicase components of tobacco mosaic virus were introduced into the tobacco genome

(Golemboski et al., 1990). The expression of the 54 k gene of TMV, the readthrough product of the 126 k and 183 k replicase, induced resistance in tobacco, and demonstrated that a non-structural gene of the virus can provide a new type of resistance. This protection effect was stronger than the coat-protein induced one, and even a high concentration of the inoculum could not break the resistance. One of the most important features of this phenomenon is that this resistance is specific either to the strain of tobacco mosaic virus from which the 54 KDa protein gene sequence was derived or to a closely related mutant, but it is not effective against related tobamoviruses. Carr and Zaitlin (1993) pointed out in their recent review that their attempt to generate resistance-breaking TMV mutants failed, which suggests that point mutations are not enough to modify a TMV strain to overcome this type of resistance. Replicase mediated resistance has now been used against the pea early-browning virus, cucumber mosaic virus and potato virus X with high efficiency, extending the possibility to control viral diseases with these different methods. An understanding of the mechanisms of replicase-mediated resistance is needed, as it is needed in the case of coat protein mediated resistance.

A novel strategy was developed by Kollár et al. (1993) to obtain virus resistance using defective interfering (DI) virus particles. This molecular parasite RNA is frequently found in association with different tombusviruses. The presence of this molecule, whose sequence is derived from the helper virus genome, results in efficient symptom attenuation and inhibition of virus replication. Engineering and cloning this RNA molecule into the plant genome has resulted in the prevention of the occurrence of the apical necrosis of the plant, including the rapid death of the inoculated plants. Extending this recent strategy to other tombusvirus-host relationships will also open alternative strategies to achieve virus resistance. Using artificially-made defective interfering plant virus particles could involve plant viruses other than tombusviruses in this technology.

Viral satellite RNAs have been found and characterized quite early, and their symptom modulating ability detected. However, several disease epidemics have been attributed to the presence of the necrotic satellite RNA of cucumber mosaic virus on tomato, while there is also attenuation of the diseases in certain cases. New virus satellite combinations have been made to attenuate the virus and to control virus diseases. In China, since 1981 thousands of hectares of different crops have been inoculated with attenuated satellite RNA as a biological agent against cucumber mosaic virus (Tien 1993). This approach was included in the transgenic arsenal by integrating the cloned satellite RNA and integrating it into the host genome. Although first reports gave evidence of the usefulness of this technique in the creation of resistance, most plant virologists express their reserves in using this approach. The risk of producing a more severe mutant satellite RNA is quite high, because a single nucleotide change is enough, and satellite RNA can associate with any helper virus, inducing even more severe diseases.

Antisense RNA molecules that bind specific messenger RNAs can selectively turn off genes. Antisense RNA of the coat protein gene of cucumber mosaic virus in transgenic plants appears to confer some protection against challenge infection.

Similar results were obtained for potato virus X and tobacco mosaic virus, but in all cases the protective effect was lower than in the case of coat protein mediated cross protection. The antisense RNA is supposed to interfere with viral replication or translation by competition for viral and host components, or by hybridizing with the sense RNA. Similar competition may be initiated with the saturation of the ribosomes with small viral sense sequences. This envisaged technology has to be confirmed experimentally.

Ribozymes, small RNA molecules, act as enzymes and cut the RNA specifically under certain structural conditions. These molecules are derived from the satellite tobacco ringspot virus, or viroids and viroid-like satellite RNAs. The cleavage is intramolecular, but the specific catalytic domain can be engineered to cleave the specific target RNA in trans. This can therefore also be used to slice the specific attacking viral RNA target. Spectacular results have been shown *in vitro*, while less progress is known *in vivo*. By engineering attenuated virus strains to amplify subgenomic RNA, a ribozyme will improve these protection technologies.

All of the above listed strategies have proved to be useful to control viral diseases, but transgenic plants continuously expressing viral sequences have called our attention to the need to evaluate the risk associated to this phenomenon. The risk of the field release of these plants has to be considered. A cloned integrated viral sequence in the plant may accelerate the evolution of viruses through recombination. To date there has been no experimental evidence that this can occur. Regarding the most advanced protection technology, there are at least three major areas of risk associated with using coat protein mediated cross protection. These are transencapsidation, synergistic interaction and recombination (Hull, 1990, de Zoeten 1991, Tepfer 1993), (for illustration see Figure 1)

Transencapsidation

Encapsidation of a viral genome in the capsid protein by another capsid protein has been shown in *in vitro* experiments with almost all the well characterized plant viruses (see Mathews 1987). In some special cases, heterologous coat protein coating was used for the protection of nucleic acids against nucleases. Heterologous encapsidation has been shown *in vivo* between virus strains, and from these experiments it has been suggested that the coat protein has an important role in insect transmissibility. Lecoq and his co-workers (1993) elegantly proved in their recent paper the aphid transmission of a non-aphid-transmissible strain of zucchini yellow mosaic potyvirus from transgenic plants expressing the coat protein of another potyvirus, namely plum pox virus. Even if this transencapsidation occurs frequently, it is a dead end for the virus, because no genetic change of the virus occurs by encapsidation of the "wrong" coat. The only danger is that transencapsidation would enable the virus to be taken along by a new vector aphid and then deposited onto a new non-host crop.

Figure 1 Major types of virus populations. A) Single virus strain population; B) Mixed virus population; C) Heteroencapsidation: D) Recombinant viruses in the population.

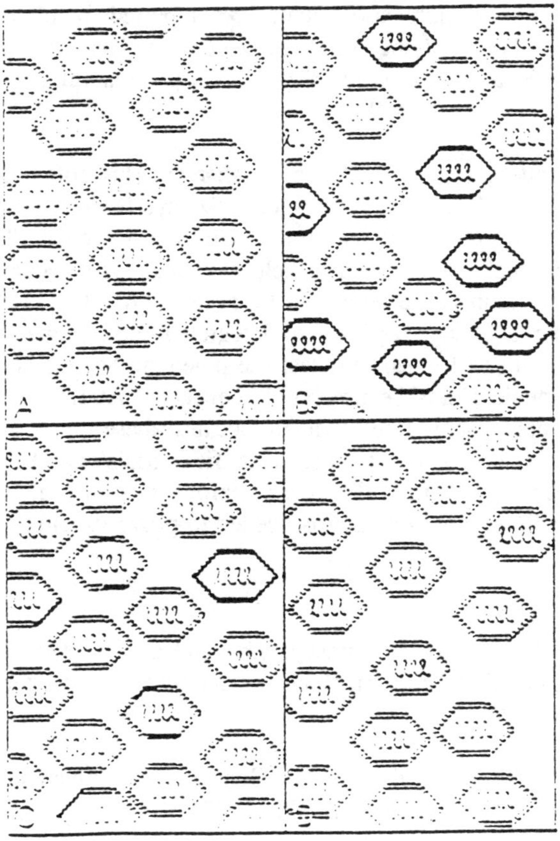

Synergistic interaction

In nature, mixed infections of plants are usually common, and viruses in several combinations cause much more severe symptoms than those induced by individual viruses. One of the most spectacular examples is the disease called corn lethal necrosis which is induced by a mixed infection of maize dwarf mosaic virus and maize chlorotic mottle virus. Indications from recent experiments that coat proteins have some role in these interactions called for this possibility to be considered, but again this has limited impact on the environment because in this case also no genetic change occurs.

Recombinations

Plants having integrated viral sequences can lead to the production of both DNA-DNA and RNA-RNA recombinations. Recombinations can take place in the cytoplasm and in the nucleus and the nucleic acid flow can happen in both directions. Homologous recombinations have been reported for plant viruses both *in vivo* and *in vitro*. By studying viral sequences it has become evident that recombination can occur in the case of plant RNA virus infection (Bujarski and Kaesberg, 1986, Cascone et al., 1993).

RNA viruses are undergoing constant genetic change. This common mechanism of change is due to the high error frequency of RNA synthesis, which has led to the acceptance of the concept that RNA viruses are "quasispecies". The advances in molecular biological techniques will also help to detect these recombination events. Since RNA recombination is the only realistic scenario for the emergence of a new virus resulting from a cloned copy of an integrated viral sequences, experiments should be designed and performed to detect the potential development of a recombinant virus in transgenic plants bearing viral sequences. At first, coat protein gene bearing transgenic plants should be involved, because these plants are already being subjected to field experiments and large scale field trials will be conducted with these plants in the near future. However, before this advanced technology is introduced in agriculture, more research is needed to prove its safety.

REFERENCES

Bar Joseph, M., Roistacher, C.N., Garnsey, S.M. and Gumpf, D.J. (1981): A review on tristeza, an ongoing threat to citriculture. Proceedings of the International Society of Citriculture 1:419-423.

Bujarski, J.J. and Kaesberg, P. (1986): Genetic recombination between RNA components of a multipartite plant virus. Nature 321:528-531.

Bevan, M.W., Mason, S.E. and Goelet, P. (1985): Expression of tobacco mosaic virus coat protein by a cauliflower mosaic virus promoter in plants transformed by Agrobacterium. EMBO J. 4:1921-1926.

Cascone, P.J., Haydar, T.F. and Simon, A.E. (1993): Sequences and structures required for recombination between virus-associated RNAs. Science 260:801-805.

Carr, J.P. and Zaitlin, M. (1993): Replicase-mediated resistance. Seminars in Virology 4:339-347.

Couzzo, M., O'Connell, K.M., Kaniewski, W., Fang, R.-X., Chua, N.H. and Tumer, N.E. (1988): Viral protection in transgenic tobacco plants expressing the cucumber mosaic virus coat protein or its antisense RNA. Bio/Technology 6:549-555.

van Dun, C.M.P. and Bol, J.F. (1988): Transgenic tobacco plants accumulating tobacco rattle virus coat protein resist infection with tobacco rattle virus and pea early browning virus. Virology 167:649-652.

Farinelli, F. and Malnoe, P. (1993): Coat protein mediated resistance to potato virus Y (PVY) in tobacco: evaluation of the resistance mechanisms. Is the transgenic coat protein required for protection ? Mol. Plant-Microbe Inter. 6:284-292.

Fehér, A., Skryabin, K.G., Balázs, E., Preiszner, J., Shulga, O.A., Zakharyev, V.M. and Dudits, D. (1992): Expression of PVX coat protein gene under the control of extensin-gene promoter confers virus resistance on transgenic potato plants. Plant Cell Rep. 11:48-52.

Gielen, J.J.L., de Haan, P., Kool, A.J., Peters, D., van Grinsven, M.Q.J.M. and Goldbach, R.W. (1991): Engineered resistance to tomato spotted wilt virus, a negative-strand RNA virus. Bio/Technology 9:1363-1367.

Golemboski, D.B., Lomonossoff, G.P. and Zaitlin, M. (1990): Plants transformed with a tobacco mosaic virus nonstructural gene sequence are resistant to the virus. Proc. Natl. Acad. Sci. USA 87:6311-6315.

Hemenway, D., Fang, R.-X., Kaniewski, W.K., Chua, N.-H. and Tumer, N.E. (1988): Analysis of the mechanism of protection in transgenic plants expressing the potato virus X coat protein or its antisense RNA. EMBO J. 7:1273-1280.

Hull, R. (1990): The use and misuse of viruses in cloning and expression in plants. in Recognition and Response in Plant-Virus Interactions. Springer-Verlag Berlin. ed. R.S.S. Fraser, p. 443-457.

Kallerhoff, J., Perez, J., Bouzoubaa, S., Bentahar, S. and Perret, J. (1990): Beet necrotic yellow vein virus coat protein-mediated protection in sugar beet (*Beta vulgaris* L.) protoplasts. Plant Cell Rep. 9:224-228.

Kawchuk, L.M., Martin, R.R. and McPherson, J. (1990): Resistance in transgenic potato expressing the potato leaf roll virus coat protein gene. Mol. Plant-Microbe Interact. 3:301-307.

Kollár, Á., Thole, V., Dalmay, T., Salamon, P. and Balázs, E. (1993): Efficient pathogen-derived resistance induced by integrated potato virus Y coat protein gene in tobacco. Biochimie 75:623-629.

Kollár, Á., Dalmay, T. and Burgyán, J. (1993): Defective interfering RNA mediated resistance against cymbidium ringspot tombusvirus in transgenic plants. Virology 193:313-318.

Lawson, C., Kaniewski, W., Haley, L., Rozman, R., Newell, C., Sanders, P. and Tumer, N.E. (1990): Engineering resistance to mixed virus infection in a commercial potato cultivar: Resistance to potato virus X and potato virus Y in transgenic Russet Burbank. Bio/Technology 8:127-134.

Lecoq, H., Ravelonandro, M., Wipf-Scheibel, C., Monsion, M., Raccah, B. and Dunez, J. (1993): Aphid transmission of a non-aphid-transmissible strain of zucchini yellow mosaic potyvirus from transgenic plants expressing the capsid protein of plum pox potyvirus. Molec. Plant-Microbe Inter. 6:403-406.

MacKenzie, D.J. and Tremaine, J.H. (1990): Transgenic Nicotiana debneyii expressing viral coat protein are resistant to potato virus S infection. J. Gen. Virol. 71:2167-2170.

MacKenzie, D.J., Tremaine, J.H. and McPherson, J. (1991): Genetically engineered resistance to potato virus S in potato cultivar Russet Burbank. Mol. Plant-Microbe Interact. 4:95-102.

Matthews, R.E.F. (1987): The changing scene in plant virology. Annu. Rev. Phytopath. 25:10-23.

McKinney, H. H. (1929): Mosaic diseases in the Canary Islands, West-Africa and Gibraltar. J. Agric. Rev. 39:557-578.

Murry, L.E., Elliott, L.G., Capitant, S.A., West, J.A., Hanson, K.K., Scarafia, L., Johnston, S., DeLuca-Flaherty, C., Nichols, S., Cunanan, D., Dietrich, P.S., Mettler, I.J., Dewald, S., Warnick, D.A., Rhodes, C., Sinibaldi, R.M. and Brunke, K.J. (1993): Transgenic corn plants expressing MDMV strain B coat protein are resistant to mixed infections of Maize Dwarf Mosaic Virus and Maize Chlorotic Mottle Virus. Bio/Technology 11:1559-1564.

Powell, P.A., Nelson, R.S., De, B., Hoffman, N., Rogers, S.G., Fraley, R.T. and Beachy, R.N. (1986): Delay of disease development in transgenic plants that express the tobacco mosaic virus coat protein gene. Science 232:738-743.

Sanders, P.R., Sammons, B., Kaniewski, W., Haley, L., Layton, J., LaVallee, B.J., Delannay, X. and Tumer, N.E. (1992): Field resistance of transgenic tomatoes expressing the tobacco mosaic virus or tomato mosaic virus coat protein genes. Phytopathol. 82:683-690.

Sanford, J.C. and Johnston, S.A. (1985): The concept of pathogen derived resistance: Deriving resistance genes from the parasite's own genome. J. Theor. Biol. 113:395-405.

Stark, D.M. and Beachy, R.N. (1989): Protection against potyvirus infection in transgenic plants: Evidence for broad spectrum resistance. Bio/Technology 7:1257-1262.

Tepfer, M. (1993): Viral genes and transgenic plants. Bio/Technology 11:1125-1129.

Tien, P. (1993): Personal communications.

Tumer, N.E., O'Connell, K.M., Nelson, R.S., Sanders, P.R., Beachy, R.N., Fraley, R.T. and Shah, D.M. (1987): Expression of alfalfa mosaic virus coat protein gene confers cross-protection in transgenic tobacco and tomato plants. EMBO J. 6:1181-1188.

van der Wilk, F., Posthumus-Lutke Willink, D., Huisman M.J., Huttinga, H. and Goldbach, R. (1991): Expression of the potato luteovirus coat protein gene in transgenic potato plants inhibits viral infection. Plant Mol. Biol. 17:431-439.

De Zoeten, G.A. (1991): Risk assessment: Do we let history repeat itself? Phytopathology 81:585-586.

GENETIC VARIATION OF VIRAL AND SATELLITE RNAS

Dr. F. García-Arenal, Laboratorio de Patología Vegetal, E.T.S.I.Agrónomos, Spain

INTRODUCTION

Plants with engineered resistance to virus-induced diseases have been obtained by the use of different viral genes, in native or modified form, or by the use of ameliorative satellite RNAs (satRNAs) (reviewed in Scholthof et al., 1993; Tepfer, 1993). Among the different strategies reported, resistance due to satRNAs, coat protein genes, and parts of replicase genes, has proved to be highly effective for very different viral systems, and has received much attention. Three main mechanisms of risk associated to the release of these genetically engineered resistant plants have been proposed (Tepfer, 1993):

1) **Genetic drift of viral and satRNA sequences due to the accumulation of mutations;**
2) **Heteroencapsidation of the genetic material of viruses infecting the transgenic plants in the coat protein expressed by those plants; and**
3) **Recombination between the genetic material of viruses infecting the transgenic plants, and viral sequences expressed in these plants.**

While 2) would be just a one-time phenomenon, mechanisms 1) and 3) may have longer term effects on the epidemiology and population genetics of the viruses to which control is aimed. RNA genomes, such as those of most plant viruses, are known to mutate frequently, and are thought to recombine only rarely (Martinez-Salas, et al., 1985; Lai, 1992), but to estimate the role of mutation and recombination in shaping the structure of viral populations data are needed on the variability and evolution of viral populations under natural conditions. Here I will briefly review our work describing the structure and variation of the populations of two plant viral systems that differ widely in their biology and ecology, and on the cautionary tales that may be derived, and that are pertinent to the use of transgenic virus-resistant plants.

APPEARANCE, SPREAD AND GENETIC VARIATION OF THE SATELLITE RNA OF CUCUMBER MOSAIC VIRUS IN EPIDEMICS IN EASTERN SPAIN

Cucumber mosaic virus (CMV) is a cosmopolitan virus of great economic importance in many crop plants, in particular in horticultural ones. CMV has isometric particles and a tripartite single-stranded (+) sense RNA genome. CMV has a very large host range, is efficiently transmitted by many aphid species in a non-persistent manner, and numerous strains have been described, indicating high phenotypic variability. Some CMV isolates encapsidated a small (340-390 nt), single-stranded RNA, that depends on CMV (helper virus) for its replication, encapsidation and transmission. This satRNA may modify in various manners (including attenuation) the symptoms induced by CMV. Also, the presence of a satRNA results in the depression of CMV replication, and has been proposed as a control agent for CMV, in cross protection or in transgenic plants, (for a recent review see Palukaiatis et al., 1992).

CMV has been endemic in all horticultural areas of Spain, including the Eastern coast, for many years. Field inspections carried on before 1986, or dry-leaf samples gathered from these areas before this date, showed CMV to be free of satRNA (our unpublished work). In 1986 a tomato necrosis syndrome appeared near Valencia, which was due to CMV plus a satRNA that is necrogenic for tomato. This syndrome spread epidemically north and south of Valencia through the East Coast until 1991; this spread may be associated with the "infection" of local CMV isolates by satRNA, or otherwise with the spread of CMV+satRNA isolates, as shown by the increase in frequency of CMV isolates with satRNA in Valencia from 1989 to 1990 and 1991 (Table 1). A collection of CMV+satRNA isolates was gathered near Valencia in the springs of 1989, 1990 and 1991. CMV+satRNA variants from these isolates were characterized by the pattern of bands in a ribonuclease protection assay (RPA) using a probe complementary to a reference satRNA (Aranda et al., 1993). RPA data showed the satRNA population to be composed of a high number of variants (the ratio haplotype in RPA/isolate being 1, Table 1), to have high values of intrapopulation diversity, and to maintain these values through the period studied (Table 1).

Table 1 Details of CMV-satRNA populations[a)]

	1989	1990	1991	Total
CMV isolates	24	22	11	57
Isolates with satRNA	12	21	9	42
satRNA variants	18	32	12	62
Haplotypes in RPA	18	30	12	60
Diversity values	0.057	0.062	0.054	

[a)] From reference 1

There was also evidence for a slow replacement of variants in the population with time (not shown). Values for nucleotide substitution per site between different satRNA variants reached a maximum value of 0.142, a value similar to that found among 23 CMV satRNA variants from different parts of the world (Fraile et al., 1991). Phylogenetic analyses showed CMV+satRNAs to evolve along a few evolutionary lines by the accumulation of mutations (Fig. 1A).

Thus, CMV+satRNA, a highly infectious molecular parasite of CMV, may spread efficiently in CMV populations, and may evolve rapidly by the accumulation of mutations to reach what appears to be an upper threshold of genetic divergence. It has been shown that this limit to divergence may be related to the maintenance of a functional secondary structure (Fraile et al., 1991). Genetic change is coupled to phenotypic change, and our field observations show that necrogenic satRNAs are substituted for less virulent forms. Nevertheless, it is important to stress that CMV+satRNA in non-necrogenic variants does not disappear from CMV isolates in the affected areas.

We do not know the origin of the satRNA epidemic, though it is quite possible that satRNA was present in solanaceous weeds such as *Nicotiana glauca*. Its spread from these undetected natural hosts may be associated with changes in the number and species composition of aphid populations in Valencia that occurred around 1986 (Jorda et al., 1992).

Thus, our data clearly show that the use of unmodified CMV+satRNA for transgenic resistance could lead to spread of satRNA into previously existing CMV populations, and give rise to new genetic types by genetic drift, which may have important potential consequences.

HIGH GENETIC STABILITY TOBACCO MILD GREEN MOSAIC VIRUS ENDEMIC IN WILD *NICOTIANA GLAUCA*

Tobacco Mild Green Mosaic Virus (TMGMV), a close relative of tobacco mosaic virus, is a plant virus with a monopartite, single stranded, (+) sense RNA genome. In contrast with CMV, TMGMV has a narrow natural host range (limited to *Nicotiana* spp.), is transmitted by direct contact between plants, and has no known natural vector. TMGMV has been reported infecting wild *Nicotiana glauca* plants in several arid areas of the world. TMGMV isolates from *N. glauca* in south east Spain were characterized by the RNase T1 fingerprints of their genomic RNA. As has been shown for CMV satRNA, the ratio of haplotypes in fingerprint/isolate, was high (0.75), but intrapopulation diversity values were small, and showed little increase with the scale of distance between sites from which the isolate came: 0.0158 for isolates within a *N. glauca* stand (distant 10-100 m), 0.0216 for isolates in a 300 km transect in Spain, and 0.0342 for isolates from south east Spain and south Australia pooled together (Rodriguez-Cerezo et al., 1991). Phylogenetic analysis for TMGMV in Spain showed a star phylogeny (Fig.1B). Thus, TMGMV shows an unusual genetic stability for an RNA virus, a property it shares with a few other plant RNA viruses (e.g. see Block et al., 1987; Rodriguez-Cerezo et al., 1989). Although negative selection may be important in limiting the genetic divergence of TMGMV, the observed genetic diversity is not compatible with a purely neutral model of molecular evolution. It could, though, be explained by a model of periodic selection, in which the (infrequent) appearance of a genome with a selectively advantageous trait would lead to the (rapid) fixation in the population of the genome carrying the advantageous trait. It is to be stressed that according to our data, and to realistic assumptions on selection coefficients, the mean time to fixation of the advantageous trait is much smaller (10^3 vs. 10^9 generations) than the mean time between episodes of appearance of an advantageous trait (Moya et al., 1993).

Fig.1 *Most parsimonious phylogenetic trees for CMV satRNA in East Spain in 1991 (A) and for 17 TMGMV isolates from SE Spain (B). A CMV satRNA from Greece (Gr), and a TMGMV isolate from Australia (AUS 8) were used as outgroup for (A) and (B), respectively. Horizontal branch lengths are to scale; vertical separation is for clarity only.*

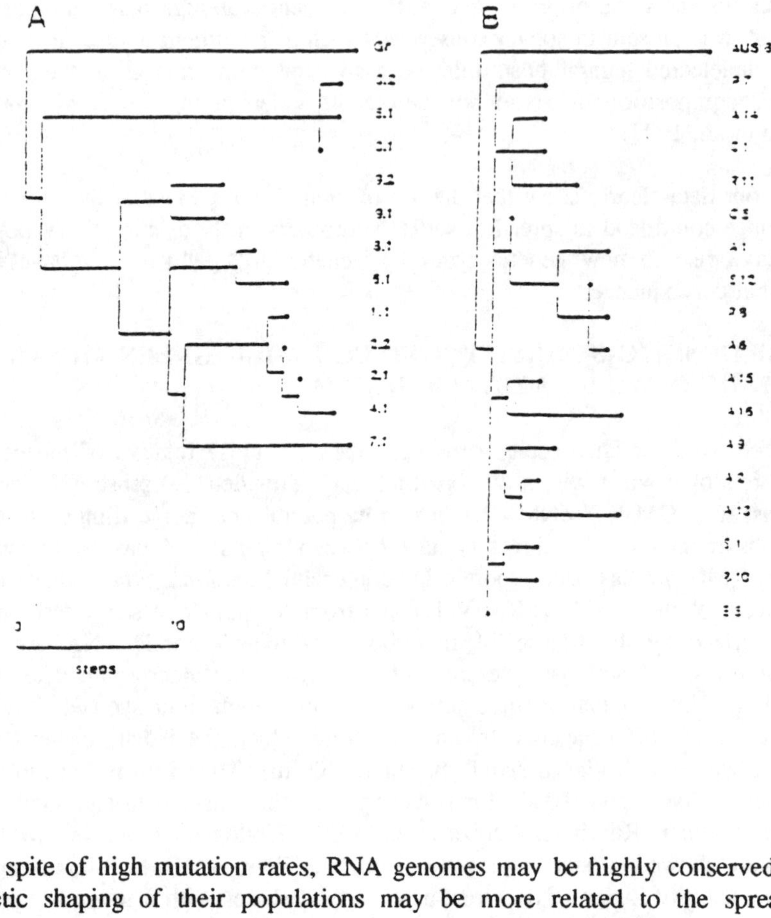

Thus, in spite of high mutation rates, RNA genomes may be highly conserved, and the genetic shaping of their populations may be more related to the spread of advantageous characters that would appear infrequently through mutation or other infrequent genetic events.

CONCLUSIONS

Our data provide evidence that great caution should be used in the use of satRNAs to produce transgenic plants resistant to viruses. They also are consistent with the hypothesis that the acquisition through recombination from transgenic plants of advantageous genetic characters could have important effects on the genetic structure (and epidemiology) of viral populations. Nonetheless, the impact in the population of a new favourable character is more strongly related to its selection coefficient, rather than to its frequency of appearance. Although the introduction of virus-derived genes into plants for engineered resistance could affect the frequency of more advantageous genomes occurring by recombination, this would not be expected to affect their selection coefficient. Thus, the conclusions that may be derived from the study of viral populations would rather suggest that pathogen-derived resistance genes can in most cases be used safely.

ACKNOWLEDGEMENTS

Work reported here was in part supported by grants PB86-0517 and AGR 90-0152, CICYT, Spain.

REFERENCES

Aranda, M.A., Fraile, A., and Garcia-Arenal, F. (1993). J. Virol. 67, 5896-5901.

Block, J., Mackenzie, A., Guy, P., and Gibbs, A. (1987). Arch. Viro l. 97, 283-295.

Fraile, A., and Garcia-Arenal, F. (1991). J. Mol. Biol. 221, 1065-1069.

Jorda, C., Alfaro, A., Aranda, M.A., Moriones, E., and Garcia-Arenal, F. (1992). Plant Dis. 76, 363-366.

Lai, M.M.C. (1992). Microbiol. Rev. 56, 61-79.

Martinez-Salas, E., Sobrino, F., de la Torre, J.C., Portela, A., Ortin, J., Lopez-Galindez, C., Perez-Brefia, P., Villanueva, N., Najera, L., Vanderpol, S., Steinhaver, D., de Polo, N., and Holland, J.J. (1985). Gene 40, 1-8.

Moya, A., Rodriguez-Cerero, E., and Garcia-Arenal, F. (1993). Mol. Biol. Evol. 10, 449-456.

Palukaitis, P., Roossinck, M.J., Dietzgen, R.G., and Francki, R.I.B. (1992). Adv. Virus Res. 41, 281-348.

Rodriguez-Cerero, E., Moya, A., and Garcia-Arenal, F. (1989). J. Virol. 63, 2198-2203.

Rodriguez-Cerero, E., Elena, S.F., Moya, A., and Garcia-Arenal, F. (1991). J. Mol. Biol. 32, 328-332.

Scholthof, K.B.G., Scholthof, H.B., Jackson, A.O. (1993). Plant Physiol. 102, 7-12.
Tepfer, M. (1993). Bio/Technology 11, 1125-1132

VIRUS RESISTANCE IN PLANTS; AN ECOLOGICAL PERSPECTIVE

Dr. J.M. Thresh, Natural Resources Institute, United Kingdom

An ecological perspective will be adopted in this paper and not that of a biotechnologist or one who has direct experience with GMOs. Attention will be drawn to the fact that plant viruses have the potential to cause devastating diseases and sometimes do so, yet in most regions and for much of the time crops and natural vegetation grow successfully and sustain only limited damage.

There are many reasons why it is important to understand the factors responsible for this apparent paradox. For example, there is a need for further increases in agricultural productivity without exacerbating the losses due to virus diseases. It is also necessary to understand the current situation and the implications of the varieties and cropping practices adopted in order to anticipate the consequences of releasing GMOs. This emphasizes the need for a comprehensive ecological approach to the complex interactions that occur between viruses, hosts and vectors within the natural and agricultural environment.

Considerable attention has been given to the nature, prevalence and susceptibility of the host as crucial factors that influence whether epidemics occur. Species that are ephemeral or sparsely distributed may evade serious disease unless conditions are exceptionally favourable for spread, even if they are highly vulnerable to infection. However, evasion alone has obvious limitations as a defence strategy and there are great advantages in some form of inherent resistance to or tolerance of infection. This leads to the concept of the 'battle of the genes' involving host resistance and pathogen virulence which can lead to the development of dynamic equilibria influenced by environmental factors that determine the overall inoculum pressure encountered.

The behaviour of many virus diseases can be interpreted in this way and serious outbreaks can be seen as major perturbations of long-established equilibria. Such disruptions can occur in various ways, as with an intensification of cropping practices or a change in weather conditions. There are also 'new encounter' situations when crops, viruses or vectors are introduced to entirely new areas or when new varieties or novel strains of virus or vector become widely distributed. Other virus disease problems have arisen following the widespread use and misuse of pesticides that has led to big increase in pest populations.

Biotechnology offers the prospect of providing a solution to such problems by greatly enhancing the ability of the host to withstand viruses and other pathogens. Some of the key issues for debate at this conference are whether such solutions are durable, whether they can be introduced without serious risk and whether they can achieve a fundamental shift in the outcome of the continuing 'battle of genes' that leads to greater stability than encountered previously. There is little definitive information on which to base even a provisional judgment but my largely intuitive response is that the capacity of viruses to cope with new challenges and constraints should not be underestimated and that problems are likely to remain with us for the foreseeable future.

** The abstract by dr. Thresh has been reproduced here since he was not able to submit his full presentation.*

PLENARY DISCUSSION AT THE END OF THE SESSION

Intervention:

Before opening the discussion, I would like to remind you of the main features that have been brought up in these three talks. Basically three major types of potential mechanisms for effects on virus populations have been identified. First, dr. Garcia-Arenal, in his talk about the importance of genetic drift, has given the best described evidence so far of this phenomenon. Genetic drift is particular worrisome if we envisage using RNA satellite genes for crop protection, since he has demonstrated that these molecules have a high rate of evolution and a high rate of drift. The second feature, on which dr. Balász has touched quite nicely, is: What is the potential for heterologous encapsidation to change how certain viruses are transmitted by vectors? Dr. Thresh pointed out also that changes in the interactions between viruses and their vectors can have potentially enormous effects on the epidemiology of virus diseases. And finally, the third point as addressed by dr. Balazs, perhaps the most knotty problem of them all, is that of genetic recombination. Under experimental conditions it has been shown that viruses during replication can simply switch templates from one RNA molecule to another. So, the question that arises is: If we have transgenic plants cultivated on a large scale synthesizing a messenger RNA, will infecting viruses simply engage in more frequent template switching between the infecting viral genome and the messenger RNA synthesized by the plant? As dr. Balázs mentioned, there is good evidence from sequencing of existing, natural viruses that this type of exchange can occur.

We may now run the risk of being accused of behaving just like the ecologists, which is to say that everything appears to be quite catastrophic, but that one should not worry too much about it. This is a little bit of a provocation, but I think that there many valid reasons to temper the way we react to these potential problems. One of these is: It is very common for plants to be infected by more than one virus at the same time. So, under these circumstances there has already been a real potential for exchange between viral genomes by template switching in plants that are co-infected. So, in a sense you can repose the question in somewhat different way concerning genetic recombination. If we have transgenic plants expressing viral sequences, is this going to lead to a higher rate of recombinations than we have already observed in the case of co-infection by several viruses at the same time? I think it is also important to recognize that these are natural phenomena. We need to be precise in order to evaluate in which cases these potential risks are really of importance.

Question:

First a comment; we have talked entirely about crops and what happens to crops, and I missed that nothing was said about the transfer of viral genes to wild relatives since virus resistance can have an ecological advantage in some habitats. I don't know whether there are some data, but there is a great possibility for such an event to occur in the genus *Beta*. The other question concerns the title of the Conference: Potential long-term ecological impacts of GMOs. I would like to hear something about the really long-term effects, not about effects in a couple of years. I think we should have a little bit more fantasy about what could happen in the long run.

Response:
We did our best and I would like to stress that we have to deal with the knowledge as it exists today. Also in this session there are areas of interest we did not have time to discuss. We tried to set the stage, in terms of trying to think about mechanisms of potential change and trying to link this to ideas from epidemiology. I think we have only tried to say: Well, here are the types of questions we should be able to try to think about; but we are not able to provide you with answers yet. Maybe after having had more discussions we may be more able to characterize long-term effects.

Intervention:
I am not quite sure whether I can provide fantasy but I can provide prejudice. I would like to point out that virology is a remarkably young science and for much of its period it has been preoccupied -and for good reason- with biochemistry, so that virus ecology and epidemiology is a grossly neglected area. So, although we know quite a lot about wild relatives and wild plants as sources of infection, we know very little about the ecological behaviour of viruses in natural communities, almost nothing at all. There is some interesting work on the viruses and vectors of maize ancestral species in Central and South America. What we do know of course is that wild collections will provide resistance genes. For example, the well known resistance gene in barley came from the Ethiopian highlands. The ecological relationship between the distribution of the resistance gene with the altitude and the exposure to infection there is a very interesting example. Wild beets are known as wild sources of infections. So, we cannot provide ecological information because there have been a very few virus ecologists and the ones that were there have been preoccupied with crops.

Question:
You have drawn attention to some potential risks associated with the use of viral transgenes. I want to make the point that one can manipulate these transgenes to effect biological containment. We have had the point about hetero-encapsidation raised. The potential risk with hetero-encapsidation, as it was pointed out, is that it may change the vector-specificity of the superinfecting virus. If one looks at the coat protein, the evidence is that the regions which control the interactions with the vector are separate from those which give the protection against the virus, and one can separate these domains and make a coat protein which would be non-transmissible and yet give protection. This could perhaps even pertain to sequences which recombine if we understood further about recombinations between viruses and one could get the potential recombination-levels down to what happens in the natural situation. The point I am trying to make is that the more we know about the details of the interactions involved in how viruses function, the safer -and I hate to use this word- the safer we can make the transgenes.

Intervention:
I am glad that you brought this up. There are at least three different cases where essentially disarmed coat protein genes have been shown to confer virus protection: Cucumber Mosaic Virus, Potyviruses and, Potato Leafroll Virus. So, here we have a way to reduce the risk. How this could be done with recombination is probably much more tricky.

Question:
Does the panel have any comment on the really long-term questions like, for instance, the degree to which plant virus virulence would change through evolution. Other groups have hypotheses about this, such as the animal parasitologists who allege that virulence should decline with evolution: Everything should become mutually benign, or some argue that virulence should at least evolve towards intermediate levels. Do the plant virologists have a view on this?

Response:
I suspect I am not enough of a virologist, but even if there is a tendency for parasitic relations to move towards a kind of mutualism, we are putting new plants, new viruses and of new vectors into new ecosystems at such an incredible rate that the long term effects can hardly be envisaged by us. What you are saying may be theoretically correct, but the cases presented by dr. Thresh, clearly show that we are modifying so much right now, in terms of new agronomic practices, that we are a long way from this type of natural re-equilibration being able to occur.

Response:
There is no simple answer to the question that you posed. That is very much depending on the particular host-virus combination that you are considering. And I would certainly say that is true of plant viruses. Admittedly, horticultural crops that are propagated vegetatively are perhaps a special case but there it is well known that for obvious reasons the horticulturalist tends to select out viruses that are relatively benign. And if we are talking about chrysanthemums, carnations and even apples and pears, then there are multiple viruses, and as dr. Tepfer pointed out, often in quite complex combinations, but where there has been a long term trend established it has been to relatively mild effects. So, if you eliminate those viruses, the benefits are really quite modest. The problems arise, if you have got an intruder like plum pox that suddenly appears and starts upsetting this equilibrium. So, admittedly vegetative propagation is not something that occurs or has not occurred in the past with vertebrate animals but certainly it is a special case. Just to make another point; necrosis is not a good thing for either the virus or for a vector that is trying to feed on a necrotic plant. So, there must inevitably be a trend away from any type of severe necrosis and, just to illustrate that point, the lettuce necrotic yellows example that I showed as my first slide is in fact a dead-end virus that is moving into lettuce entirely from wild relatives, and there is no spread from lettuce to lettuce. If there had been, necrosis would not have been an option for the virus-host pathosystem.

Response:
There is an interesting example of natural evolution of a virus in a wild plant in relation to an increase of its virulence; a certain mosaic virus that infects a wild legume in Australia makes a very bright mosaic and the rabbits don't eat the plants that have the mosaic. It increases the survival of the plant. So, this is the other way around; things are quite complex sometimes.

Intervention:
Just a comment on viruses in wild hosts; I know a couple of examples. When I go into the field I can isolate peanut stunt virus, for example, in any white clover, but I cannot see any symptoms of being infected by it. But that virus is then moved by a vector into a crop plant and depending on into which plant it goes, the effects may be quite devastating. So, I think there might be some cases where we could look at an ecological equilibrium where you got an infestation-reservoir of a virus in a wild plant.

Intervention:
I want to come back to an earlier intervention. Although I don't know anything about viruses, I know something about fungi. In the past there was a fair equilibrium between crop host plants and several parasitic fungi on these crop host plants. The point I would like to make is that virulence is not a characteristic of the pathogen, be it a virus or a fungus, it is a characteristic of the interaction between the pathogen, the host plant and even the vector. What happened in the fungal area was that host plants were selected so that the fungus became more virulent. In fact the change was not in the fungus but in the host plant. New breeding strategies aimed at reintroducing the old fashioned and nearly lost type of resistance which we now call partial resistance so that the fungi became less virulent. I think, but we have no proof yet, that the same could be done with virus resistance. We could probably select host plants that are less susceptible so that the virulence of the virus decreases. So, there is a strategy to meet today's unbalance since this unbalance between a pathogen and a host is a temporary phenomenon.

Response:
Your point about the nature of virulence being a feature of an interaction is absolutely correct. On the other hand, I am not sure whether for many cultivated plants the potential for partial resistance is available. It is not central to my own work, but in talking with colleagues who are working together with plant breeders I have got the feeling that they are often actively attempting what you just described, to reintroduce ensembles of partial resistance genes in order to create better protection. But there are many examples where these potential genes for partial resistance are not available. So, I think it is a little bit difficult to generalize.

Intervention:
The increased virulence of fungal parasites is, as you just said, not a trait associated with crops. There are a number studies on wild plants and fungal parasites that shows that high virulence can lead to the extinction of the population of the wild host and recolonization by different genotypes.

CONCLUDING REMARKS ON TRANSGENIC PLANTS EXPRESSING VIRAL GENES

Dr. M. Tepfer (rapporteur), Laboratoire de Biologie Cellulaire, Institut Nationale de Recherche Agronomique, France

One of the major objectives of plant breeding during the past several decades has been the introduction of pathogen resistance genes into cultivated varieties. These efforts have contributed to rendering agriculture both more productive and more ecologically sound, since pesticide use can be decreased if resistance genes are available. In spite of overproduction in Europe and in certain other areas, there is still a need for increased food production from a world perspective, and novel virus resistance genes will continue to play an important role in plant breeding. In many cases, natural virus resistance genes are not known, and artificial virus-derived genes will be of use not only to increase productivity, but also to decrease use of pesticides (nematicides, fungicides and insecticides) currently used as a means of eliminating the vectors of viral pathogens.

An increasing number of artificial virus resistance genes are being tested in numerous labs worldwide. These include genes encoding viral coat proteins, viral movement proteins, viral replicase proteins, viral non-coding RNAs, as well as virus-associated satellite RNAs (see Balázs, this volume). Though the mechanisms of resistance are as yet poorly understood, it is already clear that the different genes confer different levels and types of resistance. If we look at how virus populations have responded in the past to the agronomic use of natural virus resistance genes, the durability of protection has been extremely variable, going from two to more than fifty years. Thus it is reasonable to predict that virus populations will respond in a similarly unpredictable and variable manner to the artificial genes being developed. It was pointed out during discussions that virus populations in non-agronomic ecosystems are very little known, though one could predict that there would be a tendency for undisturbed ecosystems to reach an equilibrium, due to long-term interactions between the viruses and their hosts. It was thus suggested that human use of naturally resistant germplasm has already had large perturbing effects on virus populations. Though the new virus-derived resistance genes will provide new tools to plant breeders, we can not at this time predict that they will durable modify the nature of the "battle of the genes" between plant breeders, viruses and their vectors (Thresh, 1990, and this volume).

In addition to the above responses of virus populations to novel resistance genes, three mechanisms by which certain virus resistance genes could affect the genetic make up of virus populations have been described (Hull, 1990; Palukaitis; 1991; Tepfer, 1993). Genetic drift in satellite RNAs could lead to production of deleterious forms from beneficial ones (García-Arenal, this volume). Encapsidation of infecting viral genomes by coat protein synthesized by the plant can lead to changes in vector transmission of the infecting virus (Lecoq et al., 1993). Genetic recombination between virus-derived transcripts synthesized by the plant and an infecting virus genome could lead to novel viruses or virus strains. Overall, potential long-term effects concern increased variability in virus populations.

In some cases, sources of potential risk can simply be eliminated. This is the case where coat protein genes have been modified in order to render the protein synthesized non-interactive with vectors (for review see Tepfer, 1993). It has also been proposed to render satellite RNAs non-transmissible, although this has not yet been shown to be feasible.

Regarding long-term effects, the critical question was perceived as whether or not virus-derived resistance genes will, via the mechanisms described, lead to truly new and different types of perturbation of virus populations. It was pointed out that plants are frequently co-infected with more than one virus, a natural situation that would be favourable to both heterologous encapsidation and recombination between virus genomes. Thus it was argued that plants expressing sequences derived from viruses of which they are natural hosts would not provide the opportunity for occurrence of recombinants other than those that could arise during natural co-infection by two viruses.

However, there are cases where there is perceived risk of creating new types of recombinant viral genomes. Certain coat protein genes provide protection against more or less distantly related viruses. For instance, the coat protein gene of soybean mosaic virus (SMV) has been introduced into tobacco, a non-host plant for this virus, but where its coat protein gene confers resistance to potato virus Y and tobacco etch virus (Stark and Beachy, 1989). In such plants, the SMV sequences could recombine with infecting viruses, yielding new combinations that would not occur during co-infection.

Specific concern was also expressed regarding plants expressing sequences of viruses that are absent in a given area. For instance there would be specific potential risks associated with release of plants expressing sequences of plum pox virus (PPV) in the US, where this virus is absent, since PPV has proven to be a serious pathogen in Europe.

There was also a lively discussion of whether one needs to know the mechanism of resistance in order to evaluate risk. The dominant opinion of the experts in the field was that it would be more important to study in greater detail the interactions between plant, gene product and virus, in order to identify short-term interactions that could have long-term effects. This would be the first step in eliminating mechanisms of potential deleterious effects, while conserving the resistance conferring characters of genes. Though an exhaustive catalogue of areas of research priority was not made, it was clear that further knowledge is necessary in the areas of host-vector interaction, of the mechanisms of virus recombination, and how they could effect the evolution of virus genomes.

REFERENCES

Hull, R. (1990) The use and misuse of viruses in cloning and expression in plants. In: Recognition and response in plant-virus interactions. NATO ASI, Vol. H41. Ed. R.S.S. Fraser. Springer Verlag, Berlin and Heidelberg.

Lecoq, H., Ravelonandro, M., Wipf-Scheibel, C., Monsion, M., Raccah., B. and Dunez, J. (1993) Aphid transmission of an aphid nontransmissible strain of zucchini yellow mosaic potyvirus from transgenic plants expressing the capsid protein of plum pox potyvirus. Mol. Plant-Microbe Interact. 6: 403-406.

Palukaitis, P. (1991) Virus-mediated genetic transfer in plants. In: Risk assessment in genetic engineering. Eds. M. Levin and H. Strauss. McGraw-Hill, New York

Stark, D.M., and Beachy, R.N. (1989) Protection against potyvirus infection in transgenic plants: evidence for broad spectrum resistance. Bio/Technology 7: 1257-1262.

Tepfer, M. (1993) Viral genes and transgenic plants. what are the potential environmental risks? Bio/Technology 11: 1125-1132.

Thresh, J.M. (1990) Plant virus epidemiology: the battle of the genes. In: Recognition and response in plant-virus interactions. NATO ASI, Vol. H41. Ed. R.S.S. Fraser. Springer Verlag, Berlin and Heidelberg.

TOPIC 3

BACULOVIRUSES

BACULOVIRUSES

Prof. dr. J.E. Beringer (rapporteur), School of Biological Sciences, University of Bristol, United Kingdom

The baculoviruses are a particularly important group of viruses because some types have been shown to have potential to control certain insect pests, and some strains have been genetically modified to facilitate the production of large amounts of protein from cloned genes. The role of these viruses in controlling insect pests is receiving considerable attention because of pressures to examine economical and non-chemical means of controlling pests of agriculture and forestry. Even though at present these viruses have limited use, about one million hectares of Brazilian soybeans are sprayed annually to treat pest of soybean. In Brazil the incentive is the cost of control, which is cheaper than pesticides because the inoculum can be prepared by collecting dead insects from previous treatments and grinding them up to provide a source of virus.

The use of these viruses for biological control is facilitated and impeded by their attribute of only killing a limited number of host species. The restricted host range makes it easier to approve the use of such viruses as pesticides because the probability that they will harm insects (or indeed other organisms) other than the intended hosts is very low. However, from the agronomic point of view, a restricted host range is a problem because plants are often attacked by more than one insect pest and a wide host range is therefore needed.

For the farmer, chemical pesticides have the advantage that they have a wide host range, are quick acting, and are usually very reliable. In the reliability is an aspect of familiarity, since farmers are well aware of how and when to apply chemicals. Biological control agents, such as baculoviruses, by their nature depend upon the control agent being able to find and infect the pest species. Finding the host is usually no different from chemicals because both are sprayed on infested plants. Infection depends upon ingestion of viable virus particles and much work needs to be done to ensure that virus particles survive spraying and ultraviolet light for sufficient time to provide adequate protection of the crop. In bright sunlight survival is limited to a few days, other than for viruses which are sheltered behind leaves. Unlike chemicals, which are often carried systematically, fresh leaves are not protected, so spraying has to be closely related to the time of infestation.

One of the most important limitations on biological control methods is the rate at which the control agent kills the host organism. The quality and yield of a crop are affected by the amount that is eaten before the pest is killed. For chemicals this is typically very little after the chemicals have been applied. As a result farmers and the public have become used to fresh food that is not damaged by pests. For baculoviruses there is typically a 2-3 day period during which the host continues to feed, which corresponds to the time taken for the virus to replicate to the point where the host is killed. One of the most important aspects of recent molecular genetic research has been to develop viruses which carry genes whose expression will occur early during infection to produce molecules which will kill the host quickly. Examples which have been tested include hormones which affect development of the insect and toxins which will kill insects. The development of baculoviruses carrying a scorpion toxin gene will be discussed by Dr. Possee.

Clearly if these viruses are to have wide scale application for fresh foods, rapid kill of the pest will be needed. The need for rapid results presents a major difficulty in predicting whether baculoviruses will have an important role as biological control agents. On the positive side it is likely that the argument that baculoviruses will reduce the need for chemicals will retain its appeal, although an efficient publicity campaign by agrochemical companies demonstrating the safety of the latest pesticides could reduce public concern about the use of chemicals in food production. On the negative side, however, is the perceived problem that the public will be unhappy to eat plants sprayed with "viruses", and may well be even more concerned if the viruses in question are genetically modified. Whilst I have no doubts at all about the safety of the viruses that will be produced for commercial use, it will be interesting to see whether the public will be convinced that such products are safer than chemicals.

An important aspect of research with baculoviruses is the potential fate of such viruses once they have been released to the environment, either accidentally or as part of a planned operation to control insect pests. Experience with the use of various biological control agents has shown that they can be very effective in reducing the population of the host to a level that is compatible with reasonable crop yields. The final population of the host will then tend to fluctuate at a moderate level if the biological control agent persists. At its most efficient, a biological control agent may persist extremely well and will reduce the pest problem to a point where it is of little consequence. A good example of such control is the natural build-up of the microorganisms in the soil which control the disease "take-all" of cereals, caused by a fungus (*Gaeumannomyces graminis*). Baculoviruses typically do not provide this long term control. Whether or not it will be possible to modify the virus genetically to improve persistence remains to be seen.

Persistence provides two interesting problems with respect to the use baculovirus as pesticides, particularly if they are genetically modified. Under natural conditions the viruses are released from dead insects as aggregates in protein polyhedra. The protein provides protection from UV light and other environmental stresses which affect persistence. A number of genetically modified lines have been constructed which no longer produce the polyhedra protein. These have very limited ability to survive in the environment and as a result should be easy to assess as pesticides.
Whether viruses lacking polyhedron protein can be produced which survive sufficiently long, and are as infectious as is required remains to be demonstrated. The commercially attractive attribute of having extended persistence, perhaps through modification of the polyhedron protein, will present problems to regulators because such a change might be expected to confer advantages to the virus that might cause it to become a problem, particularly if another modification was to extend the host range.

In order to determine how great the potential risks might be it is necessary to know how much harm the released viruses could cause and how likely they were to spread and persist in the environment. The potential for causing harm can be tested fairly easily by examining a range of insect species to determine the host range of the baculovirus in question. The assumption would be that the target host was a pest and that damage to that species was not a problem. How far the organism might spread is another factor which should be considered. Perhaps the best model to use for predicting spread of baculoviruses is that of well diseases for plants for which there is already a fairly good database of information. This approach to estimating the impact of releases will be addressed by professor Zadoks.

Another aspect of release to the environment that might cause concern is that of baculoviruses being used under laboratory conditions. The most likely source of such releases is from laboratories working with the viruses to produce proteins from cloned genes. Kits are now available which utilize infection of tissue cultures to produce proteins from genes cloned into the virus genome. However, the risk to the environment from the release of such genetically modified viruses are negligible because the viruses concerned are genetically crippled and do not have the polyhedron protein.

RISK ASSESSMENT AND FIELD TRIAL WITH A GENETICALLY MODIFIED BACULOVIRUS INSECTICIDE

Dr. R.D. Possee, Institute of Virology and Environmental Microbiology, National Environment Research Council, United Kingdom

SUMMARY

Baculoviruses have been used as natural alternatives to chemical insecticides for over 100 years. They are more specific than chemical agents and can target particular pest insects without affecting other invertebrate or vertebrate species. In the control of soybean or forest pests they have been particularly efficacious; in other areas if agriculture, for example, the control of pests of fruit, they have yet to achieve such success. The principal problem is that the virus insecticide is slow in comparison with chemical agents, frequently requiring several days to kill the pest insects. In this time, the insect larvae continue to feed and damage the crop. This damage is often unacceptable and has hampered the use of baculoviruses to control many insect pests. The introduction of genetic engineering now offers the opportunity to improve baculovirus insecticides by inserting foreign genes encoding insect-specific toxins, hormones and enzymes. The expression of these genes within the virus-infected insect has been shown to reduce the time required to kill the host insect and to decrease the feeding damage on crop plants. To date, these studies have been conducted within contained, laboratory facilities. In order to confirm these results, small scale field trials are required to monitor the behaviour of these modified baculoviruses in the environment, where conditions may vary considerably. Prior to performing such a field test with a genetically modified baculovirus, safety assessment analyses must be carried out in the laboratory to enable predictions of the likely behaviour of the virus in the environment. This paper will describe a case study of the use of genetic engineering to modify a baculovirus insecticide, the assessment of its efficacy in the laboratory and associated experiments to predict its safety in the field.

The *Autographa californica* nuclear polyhedrosis virus (AcNPV) was modified by inserting a copy of the *Androctonus australis* insect specific scorpion toxin gene into the genome, under the control of a strong, very late virus gene promoter. Bioassays using *Trichoplusia ni* larvae demonstrated a significant decrease in the lethal time of the modified virus and a reduction in the feeding damage to cabbage plants by virus-infected insects. Further safety tests indicated that the host range of the virus was not affected by insertion and expression of the toxin gene. The recombinant scorpion toxin did not affect beetles which predate on lepidopteran larvae. These beetles were also not infected by the baculovirus insecticide. Vertebrate species such as mice and guinea pigs were not affected by the scorpion toxin or baculovirus. A fuller account of these and other safety tests will be given in the presentation.

INTRODUCTION

Baculoviruses are invertebrate-specific agents with a large (e.g., 130,000 nucleotides), circular DNA genome. Most baculoviruses are from insects, which has resulted in their use as biological control agents of pest populations (Cory, 1993). Many baculoviruses have proteinaceous virus occlusion bodies (polyhedra) which protect infectious virus particles in the period after spraying and before they encounter a susceptible host. The polyhedra largely comprise a single polyhedrin protein (28 kDa) which is produced in large quantities in the latter stages of the virus replication cycle (see Blissard and Rohrmann, 1990). The polyhedra of such viruses allow them to persist in the environment between successive hosts. This is an important feature of the baculovirus life cycle, due to the discontinuous nature of insect populations, particularly in temperate climates.

While baculoviruses have been used as alternatives to chemical agents to control insect pests, their principal disadvantage is the time required to kill the target insect. This may take several days. In this incubation period, the insect larva continues to feed on the crop, often causing unacceptable damage. Various approaches have been investigated to reduce this time. Wood et al. (1981) increased the virulence of *Autographa californica* nuclear polyhedrosis virus (AcNPV) by replication of the virus in the presence of 2-aminopurine. The lethal time 50 (LT_{50}) of a mutant, designated AcNPV HOB, was significantly lower than the parental virus. Fifth instar larvae infected with AcNPV HOB gained weight at a lower rate than the unmodified virus. Hughes et al. (1983) compared the time-mortality response of *Heliothis zea* to 14 isolates of *H. zea* NPV. Some isolates had significantly different LT_{50} values.

The most effective and predictable method for improving baculovirus insecticides, however, appears to be the insertion of genes encoding virus-specific toxins, hormones or enzymes into the virus genome. When the appropriate gene product is synthesized in virus-infected insect larvae, it can reduce the feeding activity of the insect and ultimately bring about premature death. Successful examples of this approach have been the use of an insect juvenile hormone esterase gene (Hammock et al., 1990), an insect-specific scorpion toxin gene (Stewart et al., 1991) and an insect-specific mite toxin gene (Tomalski and Miller, 1991). Each of these genes was inserted into the AcNPV genome. The recombinant viruses produced in each study offered significant improvements in insecticidal activity when compared with the unmodified, parental baculovirus. All tests with improved, genetically modified baculoviruses have been conducted in contained laboratory conditions. The results, while encouraging, have yet to be confirmed in field experiments. Such experiments represent a major advance and require careful planning and thorough assessment of their safety.

CONSTRUCTION OF A GENETICALLY MODIFIED BACULOVIRUS WITH AN IMPROVED PHENOTYPE

A field trial has been proposed to test the effectiveness of the recombinant AcNPV containing the scorpion toxin gene (Stewart et al., 1991). This virus (AcST-3) was constructed by inserting a copy of the *Androctonus australis* Hector insect-specific neurotoxin (AaHIT) coding region under the control of a duplicated p10 gene promoter, upstream of the polyhedrin gene in AcNPV (Figure 1). This arrangement retained the function of the polyhedrin gene and permitted the production of normal polyhedra. The toxin coding region was fused, in frame, with a copy of the AcNPV gp67 signal peptide coding sequence, to facilitate secretion of the toxin from virus-infected cells. The AaHIT acts by causing specific modifications to the Na^+ conductance of neurons, producing a presynaptic excitatory effect leading to paralysis and death. When neonate *T. ni* larvae were infected with AcST-3, they died 25% earlier than insects infected with the parental AcNPV. There was also a slight reduction in the lethal dose 50 (LD_{50}). Furthermore, feeding damage to cabbages by AcST-3-infected larvae was reduced by 50% (Stewart et al., 1991).

HOST RANGE STUDIES WITH AcST-3 AND THE SCORPION TOXIN

Host range of the virus

The improved insecticidal properties of AcST-3 suggested that previously non-permissive insect species might now be susceptible to the modified virus. This possibility was tested by challenging a variety of insect species with AcNPV or AcST-3. Given the logistical problems associated with obtaining and handling many different insects, it was not possible to perform normal bioassays to determine the LD_{50} in each species. For example, each virus requires 50 insect larvae per dose and 7 doses per bioassay. In each experiment the recombinant virus must be compared with the unmodified AcNPV at the same time. A compromise was reached where two virus doses (10^3 or 10^5 polyhedra) were used as a challenge. These are equivalent to 25 and 2500 LD_{50}'s respectively, as determined in a highly susceptible species such as *T. ni*.

Various insect species were collected in the field, returned to the laboratory and allowed to produce eggs. These were surface sterilized with formalin to inactivate other microorganisms. Neonate larvae were transferred to natural diet, or to semi-synthetic diet if this was acceptable to the species. When the insects had progressed to the second instar, they were assigned to separate groups, which were fed small portions of diet contaminated with 10^3 or 10^5 polyhedra (unmodified AcNPV or AcST-3) or water as an uninfected control. After 24 hours, those insects which had consumed all of the diet were transferred to fresh, virus-free diet and incubated until death or pupation. Deaths were diagnosed as a consequence of virus infection by simple Giemsa staining of smeared larvae to identify polyhedra and hybridization of purified DNA with radioactive DNA probes specific for the AaHIT coding region or the AcNPV polyhedrin gene. The results are summarized in Table 1. The various insect species were described as permissive for virus replication if they were readily infected with a dose of 10^3 polyhedra, semi-permissive if virus deaths occurred after infection with 10^5 polyhedra and non-permissive if no virus deaths resulted after this higher dose was given. The results showed that the host ranges of the parental AcNPV and recombinant AcST-3 were very similar. Differences in the percentages of insect deaths in a single species after infection with each virus are well within the normal experimental variation expected for this method. The insect species tested which are known to occur naturally at the proposed field site at Wytham were, with the exception of *A. gamma,* only semi-permissive or non-permissive for virus infection. *A.gamma* is an occasionally immigrant pest in the UK.

In summary, the host range data presented in Table 1 supported the conclusion that the recombinant virus, AcST-3, had a similar host range to the unmodified parental AcNPV. Its use in the environment would not constitute a significant risk to indigenous insect species.

Table 1: **Host-range of parent AcNPV and genetically modified AcST-3***

PERMISSIVE SPECIES (doses of < 10^3 PIBs)

Noctuidae	AcNPV	AcST-3
Autographa gamma (Silver Y)	100%	71%
Spodoptera exigua (Small Mottled Willow)	100%	42%
Trichoplusia ni (Cabbage looper)	100%	100%
Yponomeutidae		
Plutella xylostella (Diamond-back Moth)	100%	100%
Arctiidae		
Estigmene acrea (Salt Marsh Caterpillar)	98%	94%

SEMI-PERMISSIVE SPECIES (doses of > 10^3) e.g. % deaths at doses of 10^5 PIBs

Noctuidae	AcNPV	AcST-3
Agrotis segetum (Turnip Moth)	32%	24%
Agrotis puta (Shuttle-shaped Dart)	88%	92%
Apamea epomidion (Cluded Brindle)	33%	58%
Aporophyla nigra (Black Rustic)	90%	67%
Autographa jota (Plain Golden Y)	26%	10%
Caradrina morpheus (Mottled Rustic)	63%	13%
Ceramica pisi (Broom Moth)	22%	18%
Colocasia coryli (Nut tree Tussock)	10%	60%
Diarsia mendica (Ingrailed Clay)	2%	2%
Heliothis armigera (American Bollworm)	67%	28%
Heliothis zea (Cotton Bollworm)	77%	25%
Lacanobia w-latinum (Light Brocade)	69%	37%
Mamestra brassicae (Cabbage Moth)	63%	2%
Noctua pronuba (Large Yellow U'wing)	30%	33%
Noctua janthina (L.B.B.Y. U'Wing)	36%	7%
Mythimna separata (Rice Army Worm)	23%	14%
Panolis flammea (One beauty Moth)	42%	14%
Orthosia stabilis (Common Quaker)	45%	30%

Rusina ferruginea (Brown Rustic)	33%	10%
Spodoptera frugiperda (Fall Armyworm)	88%	66%
Spodoptera littoralis (Mediter. Brocade)	12%	NP
Geometridae		
Idaea aversata (Riband Wave)	4%	NP
Sphingidae		
Laothoe populi (Poplar Hawk Moth)	NP	12%
Mimas tilae (Lime Hawk Moth)	8%	92%
Nymphalidae		
Melitaea cinxia (Glanville Fritillary)	80%	80%
Polygonia c-alba (Comma)	15%	5%
NON-PERMISSIVE SPECIES (doses of > 10^5 PIBs)		
Noctuidae	AcPNV	AcST-3
Acronicta megacephala	NP	NP
Acronicta rumicis (Knot Grass)	NP	NP
Dicestra triflolii (The Nutmeg)	NP	NP
Hada nana (The Shears)	NP	NP
Hoplorina ambigua (Vine's Rustic)	NP	NP
Lacanobia oleracea (Bright-l. Brown-eye)	NP	NP
Ochropleura plecta (Flame shoulder)	NP	NP
Orthosia gothica (Hebrew Character)	NP	NP
Polia nebulosa (Grey Arches)	NP	NP
Xestia c-nigrum (Setaceous Hebrew)	NP	NP
Notodontidae		
Furcula furcula (Sallow kitten)	NP	NP
Geometridae		
Ourapteryx sambucaria	NP	NP
Pelurca comitata (Dark Spinach)	NP	NP
Lymantriidae		
Euproctis similis (Yellow-tail)	NP	NP
Pieridae		
Pieridae brassicae (Large White)	NP	NP
BUTTERFLIES		
Nymphalidae		
Aglais urticae (Small Tortoiseshell)	NP	NP

**Only data for species tested simultaneously with both viruses are included, specifically focusing on those species that earlier tests had indicated were likely to* *be infected at some dose of AcNPV (plus some other non-permissive species, as controls, and previously indicated as not infectible by AcNPV). Not shown are the uniformly negative results of tests involving non-lepidopteran insects (bees, ants, beetles, lacewings, ladybirds, sawflies etc). Other lepidoptera and other non-lepidoptera insects (in sum >70 other species) that previous studies indicated were not infectible by AcNPV were not tested. In these tests, second instar larvae were fed virus polyhedra (PIBs) in a 24 h period and then placed onto virus-free diet. The larvae were observed until death or pupation. The % mortalities at the indicated doses for virus-confirmed infections are given. NP indicates that significant numbers of insects (<2%) were not susceptible to virus infection.While subjective, experience indicates that the likelihood of any "semi-permissive" species attaining high LD_{50} doses under field conditions is, at best, remote. The likelihood of such infections leading to epizootics in those species is considered to be improbable. Semi-permissive species, like the permissive species, may contribute to virus maintenance, although for such semi-permissive species there is no evidence to support this view.*

Host range of the toxin and route of application

Injecting purified AaHIT into various insect species has revealed considerable variation in toxicity (De Dianous et al., 1987). These LD_{50} data are summarized in Table 2. Lepidopteran species, such as *Spodoptera littoralis*, were about 650-fold less sensitive than a Dipteran (*Musca domesticus*). The relative lack of sensitivity (Table 3) of AaHIT by injection in lepidopterous larvae representing some 6 species has recently been shown to be a consequence of nonspecific (ineffective) binding and proteolytic degradation of AaHIT. It is not due to a reduction in the high affinity binding to neurons since the toxin binds efficiently to lepidopterous neuronal preparations (Herrmann et al., 1990). In our own experiments, no lethal effect was recorded after injection of 300 ng of the natural toxin into 3rd instar *T. ni* larvae.

The route of application of AaHIT to insects also affects toxicity. The toxin was 500-fold less effective after topical application to *M. domestica* than by injection (De Dianous et al., 1988). Our own studies have shown that feeding 340 ng of toxin to 1 mg second instar *T. ni* larvae produced no obvious toxic effect. All larvae continued to develop normally and eventually pupated. This equates to an LD_{50} of > 35,000 ng/100mg.

Table 2: **Responses of various species to AaHIT***

Test animal (route)	LD_{50} ng/100 ml	95% confid.
DIPTERA (injection)		
Musca domestica	2	1.3-2.4
Sorcophaga argyrostoma	15	13.7-17.2
DICTYOPTERA (injection)		
Blatella germanica	26	24.3-28.0
Periplaneta americana	46	36.5-57.5

ORTHOPTERA (injection)		
Gryllus domesticus	289	224-373
Gryllus domesticus	375	384-400
LEPIDOPTERA (injection)		
Spodoptera littoralis	1310	1180-1460
MOUSE		
(sub-cutaneous injection)	> 50 mg/kg	
(intra-cerebral injection)	> 2.5 mg/kg	

**Data summarized from De Dianous et al. (1987)*

Table 3: **Responses of lepidopterous larvae to AaHIT[1].**

Insect	Dose ug/100 mg	Response*
Spodoptera littoralis	0.65	0/5
	2.0	5/5
Heliothis peltigera	2.5	0/3
	5.0	1/3
	10.0	3/3
Galleria mellonella	0.8	1/3
Bombyx mori	0.35	2/3
	0.65	5/5
Ocnogyna loewi	0.65	0/3
Cydia pomonella	0.8	0/3

[1] *Data summarized from Hermann et al., (1990).*
* *Lethality was determined by inability of insect to pupate.*

These results support the conclusion that for the recombinant toxin to have an effect on insects, it must be synthesized in vivo by the baculovirus.

Effect of the toxin on carabid beetles

After release into the environment, *T. ni* larvae are susceptible to predation by beetles. The effect of the scorpion toxin on predatory beetles was assessed by feeding virus-infected larvae to *Pterostichus madidus* in the laboratory. Adult beetles were collected from the field and fed on insect larvae until required for experimental use. They were observed to be active hunters, rapidly locating any live material within their containers. They were kept separately to prevent fighting.

The virus-infected larvae were prepared by droplet-feeding neonate *T. ni* larvae with AcST-3 or AcNPV ($2x10^6$ polyhedra/ml). Two days later, 40 fully developed first instar larvae were fed to each beetle. The beetles were monitored 1 and 20 hours after feeding, but demonstrated no abnormal behaviour and continued to hunt. The beetles were kept for a further 10 days but remained completely normal.

To quantify the amount of toxin within the virus-infected larvae fed to the beetles, 40 larvae at 75 hours post-infection were homogenized and dilutions assayed for toxin activity by dorso-lateral injection of *Musca domestica* (adult flies). This species is highly susceptible to the toxin after injection. The response to the recombinant toxin produced in the AcST-3 infected larvae was compared with that produced after injection of known amounts of the natural toxin, purified from total venom using HPLC. It was estimated that each *T. ni* larva contained 2.25-4.5 ng toxin. Since each beetle consumed 40 larvae (28 mg), they would have received 225-450 ng toxin.

In a similar experiment 8 beetles were fed approximately 20 mg of moribund AcST-3 -infected *T. ni* larvae. The next day, a further 20 mg of dead, AcST-3 infected larvae were fed to the beetles. The beetles accepted the cadavers. The beetles were maintained for a further 10 days, hunted actively and showed no behavioural changes.

HUMAN HEALTH AND SAFETY

While the host range of the genetically modified AcST-3 was the same as that of the parental AcNPV, there remained the possibility that the recombinant toxin would have an effect on non-target species. The available data on the mode of action of the AaHIT was reviewed and additional experiments performed to investigate its safety.

The symptomology (paralysis of injected insects), low doses required and rapid onset of symptoms point to AaHIT's major mode of action being on the insect nervous system. Experiments have shown that AaHIT causes repetitive firing of the insect's motor nerves resulting in massive and uncoordinated stimulation of skeletal muscle. These experiments have been carried out with several insect species including *Musca domestica* (Loret et al., 1991), *Locusta migratoria* (Walther et al., 1976), and *Periplaneta americana* (D'Ajello et al., 1972).

The selectivity of AaHIT for the insect nervous system was demonstrated by showing that no effect was observed on similar muscle preparations from members representing the Crustacea (Rathmayer et al., 1977), Arachnida (Ruhland et al., 1977), and mammals (guinea-pig) (Tintpulver et al., 1976).

In experiments using isolated single insect nerve fibres, voltage clamp experiments were performed which showed that the repetitive firing induced by AaHIT is due to a unique modification of the sodium channel conductance of the insect neuronal membrane. It was manifested as an increase of the sodium current and a slowing of its turn-off (Lester et al., 1982; Gordon et al., 1985; Zlotkin et al., 1985).

In binding experiments using synaptosomes, membrane vesicles prepared from isolated insect nervous tissue, ^{125}I-labelled AaHIT has been shown to bind to preparations from insects (*L. migratoria*, *Gryllus bimaculatus* and *M. domestica*) but not to those of crustaceans or mammals (Teitelbaum et al., 1979; Gordon et al., 1984; 1985). The toxin binds to a single class of non-interacting binding site with high affinity (Kd=1.2-3nM) and low capacity (0.5-2.0 pmol/mg membrane protein). Comparative assays with saxitoxin (which is known to bind to the sodium channels due to its displacement by tetradotoxin [Walther et al., 1976]) shows that membrane binding capacity for both toxins is the same. The binding site on the sodium channel is different from that of other well characterized toxins. AaHIT is not displaced by veratridine, tetrodotoxin, sea anemone toxin or the alpha and beta scorpion toxins which are specific for vertebrates (Gordon et al., 1985).

RECOMBINANT VIRUS AND TOXIN ACTIVITY IN SMALL MAMMALS

Although the safety of baculoviruses in small mammals is well accepted, it was considered prudent to repeat some of these tests with the recombinant baculovirus. Three separate trials were undertaken in rats and guinea pigs.

Three groups of experimental rats were injected subcutaneously with 0.5 ml of sterile water, AcNPV (unmodified virus) or AcST-3 containing 10^6 polyhedra (equivalent to 10^4 LD_{50} in 2nd instar *Trichoplusia ni* larvae). In the subsequent 28 day test period. individual bodyweights, general appearance, feeding and drinking habits were recorded. There was no difference between the 3 groups, with all animals remaining healthy. At the end of the 28 days, animals were humanely killed (anaesthetic overdose) and necropsies performed. There were no differences in the organs in animals from each of the 3 groups. The sera derived from the animals before and after the toxicity tests did not contain antibodies to AcST-3.

Three further groups of rats were incubated with 1.0 ml of sterile water, AcNPV or AcST-3 containing 10^6 polyhedra. The animals were observed for 28 days as described above and showed no abnormal symptoms. Subsequent autopsy also confirmed that there were no differences in the appearance of the organs in animals from each experimental group. The sera derived from these animals did not contain virus-specific antibodies.

Guinea pigs were randomly assigned to 3 experimental groups as for the tests with the rats described above. The test material was applied to an area of shaved back (6 cm^2) which was divided into 4 sub-areas, 2 of which were abraded using a 26 guage needle. Each group received either 0.1 ml distilled water, 0.1 ml AcNPV or 0.1 ml AcST-3 containing 10^6 polyhedra. The treated areas were re-covered with gauze for a 12 hour period. Daily observations were recorded for individual animals over a 14 day period, namely, skin reactions-erythma and oedema, body weight, general appearance, feeding and drinking habits. There were no differences between the control and experimental groups of animals, all remained normal. On day 14 the animals were humanely killed, blood samples were taken and necropsies performed. The appearance of the organs from animals in all groups was normal. The blood samples were analyzed for antibodies to the AcST-3 virus, with negative results.

Effect of insect-specific toxin in mice

Twenty adult mice were randomly assigned to 2 groups. Mice in the first group received 0.1 ml of sterile distilled water. Mice in the second group mice received 0.1 ml containing 1 ug AaHIT. Daily observations were recorded for individual animals over a 14 day period, i.e. skin reactions-erythma and oedema, body weight, general appearance, feeding and drinking habits. There were no differences between the control and experimental groups of animals, all remained normal. At the end of the experiment the animals were humanely killed (anaesthetics overdose) and necropsies performed. The appearance of the organs from animals in all groups was normal.

In this experiment with adult mice, AaHIT (1 ug) or water were applied via the intra-nasal route. All mice remained normal throughout the trial period and showed no abnormalities in internal organs.

GENETIC EXCHANGE BETWEEN BACULOVIRUSES

The transfer of the AaHIT coding region and associated transcription regulatory elements into another baculovirus might confer the improved insecticidal properties of AcST-3 on the recipient. The potential for transfer of the viral DNA sequences into other species, however, is realistic only for recipients representing other AcNPV strains, or a limited number of other, closely-related baculoviruses, co-infecting the same cells in the same insect host. Such recombination has been demonstrated experimentally between AcNPV and *Galleria mellonella* NPV (Croizier and Quiot, 1981) and AcNPV and *Rachiplusia ou* NPV (Croizier et al., 1988). These 3 viruses are regarded as different strains of AcNPV. In addition, RoNPV/AcNPV recombinants have been isolated from a wild type stock of RoNPV using plaque purification (Smith and Summers 1980). The frequency with which recombination occurs in cells infected with two baculoviruses is difficult to determine accurately. This is particularly true for closely related baculoviruses where recombinant characterisation depends on laborious analyses of virus DNA genomes with restriction enzymes and electrophoresis. Co-transfecting insect cells with infectious AcNPV DNA and plasmid transfer vectors containing a marker gene (beta-galactosidase or polyhedrin) estimated a recombination frequency of approximately 1% (Kitts et al., 1990). In similar experiments, the AcNPV transfer vector was substituted with one derived from the *Mamestra brassicae* NPV, which shares 2% sequence similarity with AcNPV. Although there was some evidence for transient gene expression from the MbNPV transfer vector in AcNPV-infected cells, stable recombinant viruses were never obtained in the virus progeny (R.D. Possee, unpublished data).

Baculoviruses which are closely related to AcNPV have never been isolated in the UK. The likelihood of recombination between a genetically modified AcNPV and other, unrelated UK baculoviruses such as MbNPV or *Panolis flammea* NPV is extremely low. In the proposed field trial with AcST-3, coinfection of the *T. ni* larvae to be employed by additional viruses will be minimal, as both insects and plants will be introduced from clean stocks and removed at the end of the study. Netted sub-enclosures on the site will also prevent access to the AcST-3-infected *T. ni* by other insects.

FIELD TRIAL WITH A GENETICALLY MODIFIED BACULOVIRUS

The studies described above represent the results of laboratory experiments and risk assessments performed between 1990-93 with a genetically modified baculovirus containing an insect-specific scorpion toxin gene. In September 1993, a field trial was performed at the Wytham field station (Oxford University) with the recombinant baculovirus to compare its efficacy as an insecticide in comparison with the unmodified AcNPV. This trial was initiated after permission had been obtained from the appropriate UK authorities and after notification of the European Community. Further details of this trial will be published in a later report.

CONCLUSIONS

The results described in this paper support the safe use of genetically modified baculoviruses in the environment. The introduction of an insect-specific scorpion toxin gene into AcNPV does not result in alteration of the host range of the virus. The recombinant toxin has no effect on small mammals or insect predators of virus-infected larvae. The potential for recombination and transfer of the foreign gene to other, distantly related baculoviruses is very small. Furthermore, the procedures adopted for the field trial may be designed to prevent the access of other insects which may harbour baculovirus infections. In September 1993, a field trial was performed at the Wytham field station (Oxford University) with the recombinant baculovirus to compare its efficacy as an insecticide in comparison with the unmodified AcNPV. This trial was initiated after permission had been obtained from the appropriate UK authorities and after notification of the European Community. Further details of this trial will be provided in a later report to be published early in 1994.

REFERENCES

Blissard, G.W.; Rohrmann, G.F. (1990): Baculovirus diversity and molecular biology. Annual Review of Entomology 35, 127-155.

Cory, J.S. (1993): Biology and ecology of baculoviruses. BCPC Monograph No 55 Opportunities for Molecular Biology in Crop Protection 3-9.

Croizier, G and Quiot, J.M. (1981): Obtention and analysis of two genetic recombinants of baculoviruses of Lepidoptera, *Autographa california* Speyer and *Galleria mellonella* L. Annals de Virologie 132, 3-18.

Croizier, G., Croizier, L., Quiot, J.M. and Lereclus, D. (1988): Recombination of *Autographa californica* and *Rachiplusia ou* nuclear polyhedrosis viruses in *Galleria mellonella* L. Journal of General Virology 69, 177-185.

DeDianous, S., Hoarau, F. and Rochat, H. (1987): Reexamination of the specificity of the scorpion *Androctonus australis* Hector insect toxin toward arthropods. Toxicon 25, 411-417.

DeDianous, S., Carle, P.R. and Rochat, H. (1988): The effect of the mode of application on the toxicity of *Androctonous australis* Hector insect toxin. Pesticide Science, 23, 35-40.

Gordon, D., Jover, E., Couraud, F. and Zlotkin, E. (1984) The binding of the insect selective neurotoxin (AaIT) from scorpion venom to locust synaptosomal membranes. Biochim Biophys Acta 778, 349-358.

Gordon, D., Zlotkin, E. and Catterall, W.A. (1985) The binding of an insect-selective neurotoxin and saxitoxin to insect neuronal membranes. Biochim Biophys Acta 821, 130-136.

Hammock, B.D., Bonning, B.C., Possee, R.D., Hanzlik, T.N. and Maeda, S. (1990) Expression and effects of the juvenile hormone esterase in a baculovirus vector. Nature 344, 458-461.

Herrmann, R., Fishman, L. and Zlotkin, E. (1990) The tolerance of lepidopterous larvae to an insect selective neurotoxin. Insect Biochemistry 20, 625-637.

Hughes, P.R., Gettig, R.R. and McCarthy, W.J. (1983) Comparison of the time-mortality response of *Heliothis zea* to 14 isolates of Heliothis nuclear polyhedrosis virus. Journal of Invertebrate Pathology 41, 246-261.

Kitts, P.A., Ayres, D. and Possee, R.D. (1990). Linearization of baculovirus DNA enhances the recovery of recombinant expression vectors. Nucleic Acids Research, 18, 5667-5672.

Lester, D., Lazarovici, P., Pelhate, M. and Zlotkin, E. (1982) Purification, characterization and action of 2 insect toxins from the venom of the *Buthotus judaicus*. Biochim Biophys Acta 701, 370-381.

Loret, E.P., Martin-Eauclaire, M.-F., Mansuelle, P., Sampieri, F. and Granier, C; Rochat, H. (1991). An anti-insect toxin purified from the scorpion *Androctonus australis* Hector also acts on the alpha- and beta-sites of the mammalian sodium channel sequence and circular dichoism studies. Biochemistry 30, 633-640.

Rathmayer, M., Walther, Ch. and Zlotkin, E. (1977) The effect of different toxins from scorpion venom on neuromuscular transmission and nerve action potential in the crayfish. Comprehensive Biochemistry and Physiology 56c 35-38.

Ruhland, M., Zlotkin, E. and Rathmeyer, W. (1977) The effect of toxins from the venom of the scorpion *Androctonus australis* on a spider nerve-muscle preparation. Toxicon 15, 157-160.

Smith, G.E. and Summers, M.D. (1980) Restriction map of *Rachiplusia ou* and *Rachiplusia ou-Autographa californica* baculovirus recombinants. Journal of Virology 33, 311-319.

Smith, G.E., Fraser, M.J.and Summers, M.D. (1983) Molecular engineering of the *Autographa californica* nuclear polyhedrosis virus genome: deletion mutations within the polyhedrin gene. Journal of Virology, 46, 584-593.

Stewart, L.M.D., Hirst, M., Ferber, M.L., Merryweather, A.T., Cayley, P.J. and Possee, R.D. (1991) Construction of an improved baculovirus insecticide containing an insect-specific toxin gene. Nature 352, 85-88.

Teitelbaum, Z., Lazarovici, P. and Zlotkin, E. (1979) Selective binding of the scorpion venom insect toxin to insect nervous tissue. Insect Biochemistry 9, 343-346.

Tintpulver, M., Zerachia, T. and Zlotkin, E. (1976) The actions of toxins derived from scorpion venom on the ileal smooth muscle preparation. Toxicon 14, 371-377.

Tomalski, M.D. and Miller, L.K. (1991) Insect paralysis by baculovirus-mediated expression of a mite neurotoxin gene. Nature 352, 82-85.

Walther, C., Zlotkin, E. and Rathmeyer, W. (1976) Action of different toxins from the scorpion *Androctonus australis* on a locust nerve-muscle preparation. J. Insect Physiol. 22, 1187-1194.

Wood, H.A., Hughes, P.R., Johnston, L.B. and Langridge, W.H.R. (1981) Increased virulence of *Autographa californica* nuclear polyhedrosis virus by mutagenesis. Journal of Invertebrate Pathology 38, 236-241.

Zlotkin, E., Kadouri, D., Gordon, D., Palhate, M., Martin, M.F. and Rochat, H. (1985) An excitatory and a depressant insect toxin from scorpion venom both affect sodium conductance and possess a common binding site. Archives of Biochemistry and Biophysics 240, 877-887.

PLENARY DISCUSSION

Question:
Could I ask your attitude to speed of kill; does that imply that you are thinking of this more or less as an insecticide, and that the pathogen has no longer term effects on the population that you are trying to kill?

Response:
I guess what you are actually referring to is the fact the viruses when they are released can obviously go on to infect other insects in the population after that initial spraying. That is certainly a point worthwhile to consider. One of the advantages of the virus is of course once you release the virus you kill the first round of insects which you target. But then the virus is still persisting in the environment and the idea then is you actually target insects in the next generation. Of course, if you modify how the virus behaves in the insect and reduce the time required to kill the host, and therefore the yield of virus, you may well actually limit the effect on the next generation of the insects. This is something which will have to be considered in future field trials. It may be that by modifying the virus in this way we actually make it worse as an insecticide over a longer term period. Having said that, most people concerned with the use of baculovirus as an insecticide are concerned that they should work much more rapidly. So, in the first instance we are aiming to actually reduce the time required to kill the host.

Question:
I agree with you that the most important problem will be the host-range of the baculovirus. There is a conflict between industrial interests, who probably wants to broaden the host-range and the ecological risks. You have tested a lot of species infected by baculovirus in the laboratory. Will you also test these effects in field tests the next few years in relation to a monitoring programme?

Response:
You are suggesting that we should actually monitor the insects in the field around the field site for the presence of the virus. Is that right?...In the present this is not one of things we are doing because we feel that the virus has been contained on the site. Now, if in the next few years we are able to do field trials in which we have viruses sprayed in a open field, that would be part of a programme. Of course, you then will get into problems like: How many insects should you sample and over what area, as you get away from the point source the area gets bigger and bigger. To be practical, I think, to do a proper survey of that sort of phenomenon would take a vast number of people. I am not saying that perhaps we shouldn't make some attempt to monitor the insects. But I think one has to be smart about it: Is it possible to sample thousands and thousands and thousands of insects to determine whether the virus is actually present?

Question:
You said adding a foreign gene should not alter host-range. But what do you actually know about what determines host-range?

Response:
At the present time we have knowledge of one or two genes which actually modify the host-range of the virus. That is not to say we understand them all. My personal perception is that a gene which is expressed only after the virus actually started to replicate in the insect-host (and basically the only way that this gene can be expressed is after the virus has infected the insect-host) the gene on its own cannot affect the host-range.

Question:
I cannot follow the logic of that. Once a nucleic acid is released, everything has to be made and presumably the coat has an influence on host-range. So, any protein you are producing from your late promoter has the possibility of attaching to the virus-particle and therefore has presumably an equal chance to influence the host-range.

Response:
Now I see the point. The thing is that protein which you express would have to have an effect on the replication of the virus, either in the sense of the uptake of the virus into the host or in the early stages of the replication. Personally I don't see how the toxin could do that.

Question:
Neither do I. But what determines host-range?

Question:
My first question refers to toxicology: Did you do also test for sub-acute and chronic effects? The second question: How do you estimate the chance that the recombinant-virus will recombine with wildtype-virus and will acquire another host-range and/or a higher fitness?

Response:
With mice we used a range of doses of the toxins and we have found no effects either on the growth appearance of the mice or in the organs like liver, kidneys and spleen within the host. Your second question on the exchange of DNA between viruses is a good one. Basically, if the viruses are very similar and you can co-infect insects with the two viruses, you can see exchange of DNA at a percentage of about 5 %, that is with two viruses which are identical. If viruses are not identical, either they have a different host-range and the similarity between the DNA is relatively low, it becomes extremely difficult to actually recombine DNA between them. In the experiments that we have done in the laboratory we found it very, very hard to recombine DNA from one virus into a different virus. Obviously, this type of phenomena must go on in the field at some percentage, because you do get exchange of DNA between viruses; this has been shown but happens at a very low level. Now

we don't have the complete answer on this, I admit. We don't know yet whether it is possible at a very low frequency for a foreign gene such as a toxin gene to hop from one baculovirus into a different one. The point to make, though, is that for the recombination to take place the two viruses must infect the same host and be in the same cell at the same time. At the present time the overlapping of baculoviruses between insect hosts is not a well characterized phenomenon. We know the detailed host-ranges for a number of baculoviruses but not many. With two baculoviruses that we do have detailed host-range knowledge, we found that when we tried to recombine them in the same insect host we cannot force it; it does not seem to happen.

Question:
Last year there was a report from Brazil on an insect developing resistance to baculovirus. Is that a special case, or are you not concerned about resistance?

Response:
Yes, we are concerned about resistance. One thing that we never say is that insects won't become resistant to the baculovirus because I am sure that that will be found in the longer term. In the Brazilian case I think the situation there is slightly different because their strategy is to collect virus from the field each year, purify it and then spray it again the next year. And I think they have recorded a twofold increase in the LD_{50} over a period over ten years. So, the resistance development over that period in time appears to be quite low in that the insects have not become resistant in a major way. There are not many examples of resistance which have been documented. Baculoviruses can be become latent within insect hosts. There is some information coming out on that now, but it does not appear to be that frequent as far as we are aware. There is a debate whether baculoviruses maintain themselves normally in the insect population in a latent state and only become apparent when the insect population rises to a level such as there is some stress factor within the insect population which causes the virus to be 'triggered' to kill the insect host. Again, that area is not well researched.

Question:
Could you reflect on the interest of industry for this system? How much are baculoviruses currently used as opposed to chemicals? Are there any other uses of baculoviruses than insect eradication?

Response:
The biggest field of application is Brazil; over one million hectares are sprayed every year with a baculovirus-preparation, not the same one as the one we are working with here. They have been used in America from time to time. The markets, at present time, are not large. Industry is now getting more interested in this field. A number of companies in the USA and the UK is having an interest in this field. And I think that much of this interest has been prompted by the possibilities of actually modifying the viruses used in this type of technology that I described today.
The other use for the baculovirus is as an expression vector in the lab. This prompted, I think, most of the work on the insecticide-aspects because people looked

at the expression of the very late genes by the virus in the insect cell; vast amounts of these proteins were made in these cells. So, over the past ten years laboratories around the world have utilized the viruses to express foreign gene products.

Question:
What is the potential for these to escape and proliferate in the environment?

Response:
This is one of the concerns which we are looking at the present time. Most of the baculovirus expression vectors which we used lack the polyhedron gene and the polyhedron gene forms that matrix protein actually around the virus particles. Without that matrix protein the virus cannot survive in the environment. Some years ago we actually did a field experiment in which we released the virus lacking the polyhedron gene coat into the environment and it did not survive for very long. So, if the expression vectors lack the polyhedron coat protein, you can consider them as being very, very safe. With the polyhedron protein there is a possibility that the viruses might survive. Whether the amounts that might be released into the environment would be a hazard in the environment is a question which we need to look at.

Question:
You addressed the question of the possibility of mutation from the toxin to one which has greater toxicity to the same mammals by doing a comparison of the whole amino acid sequences of the different toxins. Is the property of these toxins due to the whole sequence or is there an active centre which gives these properties? And if there is an active centre, what are the degrees of homologies between those active centres?

Response:
The active centres of these toxins are unknown, as far as I am aware, but the overall structureS of the toxins which are specific for mammals and specific for the insects are very, very different. There is no common site identified at the present time. It is a good point. The comparison that we do was done over the whole of the protein sequence. Having said that, even over a very limited area of the toxin you would have to argue for a fairly large amount of changes to have any effect on the toxin itself. The other point to make is that the baculoviruses are very, very stable. One of the advantages with a virus-system is that you can do many, many experiments in the lab in a very short period of time and actually duplicate the number of replications over a very long period in the field. Basically, you could do up to 50 generations per year in the lab, or even more in a cell culture-system. In the experiments we have done the viruses have been found to be very, very stable with no mutations at all in the regions which we have looked at. Obviously, changes must take place over a very long period of time but in the experiments that we have done both the toxin genes and many other genes which we have expressed in the baculovirus genomes have remained very stable. So, we feel the chances of mutations to a toxin which might affect mammals are very, very remote.

AN EPIDEMIOLOGICAL VIEW ON THE INTRODUCTION OF GMOs AT THE THIRD TROPHIC LEVEL

Prof. dr. J.C. Zadoks, Department of Phytopathology, Agricultural University Wageningen, The Netherlands

TROPHIC LEVELS

My starting point is ecology which studies the complexities of interactions in nature and, among other things, the question 'who eats whom'. Ecologists distinguish ranks among the eaters and the eaten (Zadoks & Schein, 1979). Plants are eaten. They form the first trophic level, where trophic means dining. The second trophic level is formed by the plant eaters, man, cattle, and many other organisms of which some are harmful to crops such as rats, birds, insects, nematodes, fungi, bacteria and viruses. The third trophic level consists of those organisms which eat the plant eaters. Among these are the lion and the eagle as animals of prey, but also predating and parasitizing insects and, again, fungi, bacteria and viruses.
In fact, the picture of three trophic levels is an oversimplification because there exist long trophic lines, which weave together into a trophic web, the food web. Here, we will stay at the simple side and stick to three trophic levels.

BIOLOGICAL INVASIONS

Another chapter of ecology deals with biological invasions. A biological invasion takes place when an organism suddenly appears in a new environment which is rapidly colonized. Nearly all biological invasions are man-induced. Voluntarily or involuntarily, man transports an organism to a new environment, usually a new continent (Hengeveld, 1989; this volume).

At the first trophic level, that of the plants, such invasions are not scarce but ill-described, and I will skip these. At the second trophic level, biological invasions are better studied. Examples are the plague of humans (caused by *Pasteurella pestis*), the muskrat (*Odonthra zibethicus*), the collared dove (*Streptopelia decaocto*), all in Europe, and the killer bee, in the Americas. Because of their economic importance, biological invasions of plant pests and pathogens are well documented (Zadoks & Van den Bosch, 1994).

At the third trophic level, biological invasions do occur, usually man-induced for purposes of biological control. Unfortunately, the documentation is often poor or poorly accessible. Nevertheless, it seems that even such invasions cannot escape certain rules.

Many biological invasions are typically 'point invasions'. An organism is introduced into a new environment at a specific location or point in space, from where it spreads gradually. This spread can be described and analyzed mathematically and, often, a regularity is found (Heesterbeek & Zadoks, 1987). Frequently, the organism spreads at a constant radial expansion rate. In other words, where possible it spreads in all directions invading new territory gradually but steadily. Of course, in the beginning of the invasion the expansion has to pick up speed, whereas at the end of the invasion the expansion rate may be curbed because the environment becomes less favourable.

MATHEMATICS OF INVASIONS

Mechanistic models of biological invasions applying analytical (Van den Bosch et al., 1990) and numerical (Zawolek & Zadoks, 1992) mathematics lead to the same result, an organism starting with a point invasion spreads at constant radial expansion rate when conditions are favourable. Detailed field studies at different scales of magnitude confirm the point.

If the invading organism is a pathogen, the invasion is called an epidemic. Here we talk about the epidemiology of plant pests and diseases. We distinguish three different scales of magnitude in epidemiology (Heesterbeek & Zadoks, 1987). The zero order epidemic takes place at the micro-level, within a single field. The area involved is measured in meters and the rate of spread is expressed in centimetres per day. The first order epidemic occurs at a regional scale within a single season. It involves a large number of fields scattered over hundreds of kilometres. The rate of spread is expressed in kilometres per day. The second order epidemic occurs at the continental level and may take several years. One characteristic is the winter stop. When there is no crop on the field, the epidemic cannot progress. On the contrary, the amount of pathogenic units may diminish during the winter stop. Nevertheless, we find a constant rate of radial expansion over the years, mathematically (van den Bosch & Zadoks, 1993) and in reality (Heesterbeek & Zadoks, 1987).

THE BACULOVIRUS PROBLEM

Now we can position the problem of the genetically modified baculovirus. The baculovirus is introduced to control an insect pest which devastates a crop. Trees in a forest are seen as a perennial crop. The baculovirus behaves as an organism at the third trophic level. Our fear is that its artificial introduction into the environment may start an undesirable biological invasion.

Fortunately, we are prepared. Applying the rules, laid down in mathematical equations, we should be able to predict the velocity of radial expansion or rate of spread of the new invasion, provided we had the necessary parameters available.

At the present stage we have to hunt for such parameters, or even to make educated guesses. The situation is complicated because the genetically modified baculoviruses, in which we are interested, are disarmed. First, their survival ability is reduced. Second, they may kill so fast that normal multiplication is hampered, a form of biological containment.

There are some data available about 'primary dispersal gradients', one input to feed the equations. In the Corvallis experiment (Anonymous, 1991) a small area of cabbage infested by caterpillars (*Autographa californica*) was treated with disarmed baculovirus (AcMNPV) in order to kill them. The 'primary dispersal gradient' in this case is not a natural one, but man-induced as a result of a treatment in which dosage, characteristics of the spray droplets, local turbulence and wind play a role. The observed gradient is due to turbulent diffusion of inoculum. The data can be used as a stand-in for a real primary dispersal gradient. It provides a characteristic piece of information.

If the baculovirus is only disarmed by elimination of the polyhedron formation, co-occlusion makes that the multiplication rate of the disarmed virus is one half of the normal rate. If a toxin gene of *Bacillus thuringiensis* were introduced into the disarmed baculovirus to kill the caterpillars faster, the multiplication of the virus would be reduced to an unknown but very low level. To remain at the safe side, we might assume a one per mil reduction.

An important parameter is the overwintering of the baculovirus on plant remnants or in the soil. In the Corvallis experiment, the overwintering of virus particles was below 2 percent.

These non-zero guestimates could be entered into our equation for a second-order epidemic (Van den Bosch & Zadoks, 1993) with the expected result that the calculated rate of radial expansion of a second order epidemic due to genetically modified baculovirus is so low that it can hardly be observed. Any calculated value has to be compared with known values, which are in the order of 100 km per year, for parasitoid insects, that is at the third trophic level (Zadoks, unpublished). The spread of the genetically modified baculovirus will be, comparatively speaking, negligible if existent at all.

In normal usage the genetically modified baculovirus would be applied over large areas and not at one point in space only. For practical purposes, it makes little difference whether the man-made biological invasion is a point invasion or an area invasion since a point invasion will become an area invasion in the course of time.

WORST CASE QUESTIONS

The question may arise if a practically non-existent spread is, in the long run, 'no spread' or 'slow spread'. The equations do not answer such a question because they are deterministic equations. The survival risk seems so low that in practice this slow spread equals no spread indeed.

There is an empirical reference for large-scale application of non-modified baculovirus. In Brazil, millions of hectares of soybean are treated annually with baculovirus to control the velvet bean caterpillar (*Anticarsia gemmatalis*; Moscardi & Cerrea Ferreira, 1985; Zadoks, 1992). Every year new treatments are necessary because the baculovirus has a low overseasoning and spreading capacity. The point would be true *a fortiori* for disarmed and fast-killing baculovirus. Baculoviruses are highly specific to their target organisms. No environmental damage nor risk to humans has come to my knowledge.

In a worst case scenario one should consider reverse mutations in the putative survivors. One is the acquisition of the original polyhedrin gene. The other is the loss of fast kill. Both seem feasible in principle, but at extremely low levels of probability. This low probability level, in combination with very low levels of mutable survivors, leads to extremely high levels of improbability. In other words, the worst case risk seems negligible.

CONCLUSION

The genetically modified baculovirus is a pathogen at the third trophic level, of which there exist literally hundreds of species in nature. The normal rules of spread apply. A limited number of adequately documented field data allow to (gu)estimate the parameters to be introduced into the equations. Preliminary calculations show that the expected rate of radial expansion for a second order epidemic of genetically modified baculovirus is negligibly small.

The equations used are deterministic equations derived from epidemiological theory. They give no indication of the probability of back-mutation of the modified virus. The combination of low multiplication with low survival practically excludes this hazard. According to present day epidemiological knowledge, genetically modified baculovirus without polyhedrin genes and with one or more fast-killing genes (Bt and others) is safe to use as a pesticide.

REFERENCES

Anonymous - 1991. Corvallis Environmental Research Laboratory Extramural Project. 1989-1991 Progress Report. Field release of a genetically-altered baculovirus with a limited survival capacity.

J.A.P. Heesterbeek, J.A.P., Zadoks, J.C. - 1987. Modelling pandemics of quarantine pests and diseases; problem and perspectives. Crop Protection 6: 211-221.

Hengeveld, R. - 1989. Dynamics of biological invasions. London, Chapman and Hall. 160 pp.

Moscardi, F., Cerrea Ferreira, B.S. - 1985. Biological control of soybean caterpillars, pp 703-711. In R. Shibles (ed.): World soybean Research Conference III: Proceedings. Boulder (Colorado), Westview Press.

Van den Bosch, F., M.A. Verhaar, A.A.M. Buiel, W. Hoogkamer, J.C. Zadoks - 1990. Focus expansion in plant disease. IV: Expansion rates in mixtures of resistant and susceptible hosts. Phytopathology 80: 598-602.

Van den Bosch, F., Zadoks, J.C., Metz, J.A.J. - 1993. Continental expansion of plant disease: A survey of some recent results. Predictability and Nonlinear Modelling in Natural Sciences and Economics. Proceedings of the 75th Anniversary Conference of Wageningen Agricultural University. Series: Proceedings in nonlinear Science, AV Holden (ed.); Wiley. In press.

Zadoks, J.C. - 1992. The costs of change in plant protection. Journal of Plant Protection in the Tropics 9: 151-159.

Zadoks, J.C., Schein, R.D. - 1979. Epidemiology and plant disease management. Oxford University Press, New York. 427 pp.

Zadoks, J.C., van den Bosch, F. - 1994. On spread of plant disease. A theory on foci. Annual Review of Phytopathology 32: submitted.

Zawolek, M.W., Zadoks, J.C. - 1992. Studies in focus development: An optimum theorem for the dual dispersal of plant pathogens. Phytopathology 82: 1288-1297.

PLENARY DISCUSSION

Question:
You have told us that you cannot see any environmental risk of baculoviruses. In order to avoid a message like: We don't know anything but don't worry, I would like to know what do you actually know about the ecological role of baculoviruses in natural and semi-natural ecosystems?

Response:
I know very little because that is not my area. Maybe the former speaker can answer this question better than I can. What we know in nature is that baculoviruses exist and that they are often dangerous in contained situations like in silkworm and so on and that there are outbreaks once in a while, usually very restricted. So, baculovirus is a part of nature but not a conspicuous part of nature. The reasons why it usually goes hiding and only rarely flares up we really don't know. So, we believe that the baculoviruses are rather harmless elements in nature; there are hundreds of species known, they are highly specific. The Brazilian example shows that you cannot really have the baculovirus to stay there; you have to apply it every year again, they do shift fields but even from one old field to the next new field baculovirus hardly makes that step of only a few metres. So, whatever we know from baculovirus from field observations -wildtypes- is very comforting.

Question:
Are you in a position to refute the suggestion that baculoviruses are rare because they are extremely important; and the reason that they are rare is that they attack insects that are rare in consequence?

Response:
That is a deep one. There are indeed cases where the second trophic level is very rare indeed because the third trophic level is highly effective and this is mainly the case, as far as I know, when the third trophic level consists of parasitoids with very high searching capacity; they can search the insects at the second trophic level. Now baculovirus has no searching capacity at all, it is completely passively spread. So, there must be different reasons why outbreaks of baculovirus in nature are relatively scarce. In fact I do not follow your suggestion.

Intervention:
I just want to make one comment. I think it is very interesting to apply Murphy's Law to ecology: If it can happen, it will happen. I would like to draw a distinction in this. I think there is a difference between what can happen in terms of what you can imagine and what can happen out in nature. And I think that there may be a tendency for us to be able to propose mechanisms which essentially don't exist in nature.

Response:
Thank you for that remark. The problem is I can only predict the past. I cannot predict the future. So, we have to draw our lessons from the past. In that respect I would say the outlook of crop protectionists is very, very gloomy. There are highly unexpected transfer mechanisms. We know for example that aeroplanes and stratospheric winds have transported pathogens. These are hardly things we could think off at the time sitting behind our desk. So, I really believe that Murphy's Law is valid, if we look over a century. However, the question is: Can it really happen? And that question is up to the molecular biologists to answer, and not up to the ecologists.

Question:
I invite the speaker to elaborate a bit more on his conclusion of non-equilibrium. We have a number of on term non-equilibrium phenomena. Firstly, the human population is not an equilibrium but still expanding for a few decades yet. Secondly, and more optimistically, the acquisition of knowledge and the ability to store and to diffuse it is not on an equilibrium but on continuous monotonically increasing gradient; virologists are developing the world virus data bank, the molecular biologists are giving us steadily improving methods of diagnosis and this is very relevant to the International Office of Epizootics working on the improvement of diagnostic decision-making which could enormously enhance the speed and efficacy of the control of disease. If you listened to dr. Thresh, you get this impression of continuous struggle and cycle and dance of a never-ending oscillatory equilibrium. Whereas your suggestion of a non-equilibrium situation would break out of that. It seems to me that we have some quite optimistic non-equilibrium phenomena of which, I said, the acquisition of knowledge and the improvement of, or at least the quality of, administration are important. In Africa, for example, the problems are not of science and technology, the problems are quality of administration. But in the developed world you certainly have a continuous improvement in the management and application of your biological knowledge. The question is whether you take that in the form of increased population or in the form of increased security. So far, we have taken, the last couple of centuries, improvement in management just by population increase as result of progress in immunology and public health but we could take it in the form of increased security. I come back to my starting point: Could you elaborate a little bit on what you see as non-equilibrium phenomena?

Response:
There is a technical point and there is a philosophical point in your comments. The technical point is that if genetically modified baculovirus is just another ordinary pesticide, then we will get the pesticides Boom-and-Bust-Cycle; a pesticide works for a certain number of years and then the pest becomes resistant to the pesticide and the pesticide is no longer effective. Yesterday I talked about a Boom-and-Bust-Cycle in resistance breeding, here we have a Boom-and-Bust-Cycle in pesticides. We are very used with that; we get undesirable outbreaks because a pesticide suddenly does not work anymore and we don't have another pesticide readily available. It always takes a few weeks to import it. So, that is the lack of equilibrium I wanted to indicate. If I may enter into the philosophical discussion, there are a few elements on

which I may agree with you. Human population is indeed growing fast and needs to be fed; knowledge is increasing not linearly but exponentially at a very fast rate but the trouble is that the knowledge is not applied appropriately. So, even though a lot of knowledge which we as scientists think to be to the benefit of humans is available, it is just not applied. One of the reasons may be the hesitation by the regulatory authorities but probably that hesitation is very justified; that is the discussion of this conference. Usually, knowledge is not being applied because of disorganization and you rightly mentioned the African problem where disorganization at all strata of society -not just government- is predominant. However, if you look at the South East Asian societies, you see that they are getting pretty well organized and they are rapidly taking over the Western societies nowadays. The turning point might be around 2000. So, I would not say that there is a non-equilibrium in the human society in the application of knowledge but there are certainly regional instabilities which do not allow the application of knowledge and there are institutional inhibitions all over the world. In the Western world these are even stronger than in the Asian world.

Question:
You are apparently quite relaxed about the application of baculoviruses. As I see it, this relaxation is based upon the experience being made by spraying millions of hectares without any problems. Would you agree with me that this relaxation is based upon several parameters and one of them is the fact that baculoviruses have a very specific host-range? Does this mean once you start changing the host-range specificity that your relaxation would disappear?

Response:
I could imagine that an industry wanting to invest in baculovirus would think it more profitable to expand the host-range. Let us say, for example, in cotton there are more grazing insects than one and it would be very attractive to industry -but I don't know whether it is possible- to expand the host-range to the six caterpillar species which damage cotton because you will then be selling tremendously increased pesticidal capacity. This is not yet the case. At the moment baculoviruses are highly specific and that is the major safeguard. Whenever you start applying baculoviruses that have a wider host-range, then of course you don't know the limits of the host-range and we will loose our grip on the matter. However, when you add Bt or scorpion genes combined with some biological containment, I feel confident. At the moment it is mainly the material in which you pack the baculovirus -glycerol- which is damaging to the eyes of the applicator. Maybe there is some allergenic effect to the applicator but that has not yet been demonstrated. So, it indeed seems a very safe pesticide.

Question:
Your last comment about using different baculoviruses or expanding the host-range prompted me to ask: If on a particular crop plant you may have three or four different caterpillars you want to control, I wonder if tests have been done where you have inoculated or tested mixtures of baculoviruses on a mixed population of insects and wonder what the results would be then in terms of host-range.

Response:
Well, that is a lot of "if's". I really don't know. I would say the only danger I see is that if one particular insect ingests simultaneously two different baculoviruses and that one cell of that insect is simultaneously invaded by these two different baculoviruses, then there might be some exchange genetic material. That is the only danger I see. I have to ask at the former speaker whether this could have serious consequences, for example; expansion of the host-range.

Intervention:
I think the only occasion I know of which two different viruses were used against two pests of the same tree in this instance was in Scotland and that worked very well. Nobody did any studies on exchange of information between those two viruses because these were wildtype viruses and to actually do those studies would have been extremely difficult.

Question:
The host-range is important and we have to do testing and sampling in each phase but that will be difficult. That is what I have understood so far. One of the blocks we have as regulators is that we would like to offer people carrying out those field tests is the kind of data and monitoring they have to provide to us. Can we reach some kind of agreement among ecologists on what kind of tests should be done; how many species you have to sample, their numbers, the area and so forth? Will it ever be possible to design a monitoring programme on which you can base a well balanced decision?

Intervention:
With the host-range tests that we have done prior to the field release experiments, we have always worked on the basis that we should collect insects from the field. So, we are not working with insects that have been reared in the lab for years and years. So, we are working as close as possible to the natural situation. Testing on that level is relatively simple because it just involves collecting insects from the fields and it does not matter where you collect them from provided that they are from the wild. You then can rear those insects for a brief period in the lab and collect eggs, hatch them out and use those larvae in the host-range tests. That is relatively simple and could be done for a long period of time depending on the number of insects which you can collect from the field and on the seasons in which you can collect the insects. As far as monitoring after a field experiments looking for the presence of viruses in the natural populations of the insects concerns, that is a different kettle of fish. It would be nice to say: Yes, we could sample over a vast area, but on a practical level how do you do it. The further you get away from your field release point, the area gets bigger and bigger and it gets harder and harder to find insects and if you do find the insects, what kind of tests should you do on them; how do you monitor the presence of the virus? PCR is possible and that would mean doing vast numbers of PCR reactions; but what then does a positive result mean? So, monitoring after a field release experiment is actually very difficult to tackle.

Intervention:

Yes, it is probably theoretically possible to design a sampling strategy but as has just been said the cost are absolutely going to be enormous. It may well be possible if you were to do some more fundamental ecology of the baculoviruses. You may then get some sort of vague idea about how much gene-exchange there was naturally. But again, I would not like to do that. I mean: If you look at the natural history of baculoviruses using genetical markers, that is going to be more likely to give you some interesting information than actually trying to do some sort of direct experiment. To do that kind of monitoring is going to cost far too much money.

Response:

I agree with the last two speakers. If you want to prove that it does not happen, just allow a small experiment and you will not find it. If you want to prove that it happens, you need to do an experiment over hundreds of hectares with highly infested crop plants; the chance you will find it can be compared with finding a needle in a haystack and even if you use instead of PCR bio-assays, which seems simpler to me, then even so you will not find it probably. So, the only way to go about has been indicated by the last speaker: Try to know the evolutionary history of baculoviruses! If you want to fund that research, you should have at least 10 million ECU's in your pocket, then maybe you get a first sketch of an answer.

Question:

I heard two arguments. The first one; it is going to be very difficult and expensive. Actually, I am not interested in that. If it is expensive or not, I do not care. If we need the data, we have to ask them and if they cannot provide these, that does mean stop. The question is: Do we need those monitoring data? Before we force somebody to really go into those kind of monitoring problems, we need a reason to do so. My first impression would be that if based on laboratory tests carried out as described by dr. Possee there is no indication to believe there is a broader host-range, then probably you should not even ask for further monitoring.

Intervention:

Lab tests on non-target organisms are obviously excellent but I think the issue has been coming up now tangibly for a couple of times. People have been worried about field trials or monitoring of impacts on non-target organisms. I suggest you from a statistical point of view that it is not just expensive, it is also impossible to do sensible monitoring on non-target organisms. Where for example would your controls be? You would measure densities of insects around your field experiments but you will have no way of attributing those fluctuations or those densities to the fact that the baculovirus got out. You would have to do independent experiments where you yourself control the densities of non-targets before and after you have infected them. So, it is not just a waste of money, it would also be a waste of science because you will not be able to come up with statistically significant convincing results.

Question:
I wonder whether it would be possible to make a sort of single shotgun from this virus by attaching the toxin gene to an early promoter and so you could eliminate the appearance of adult viruses and you would not need to worry about changes in the host-range. Is this technically possible?

Intervention:
Yes, but it is difficult because if you express the toxin gene under a promoter active in the early phase of the replication cycle, you may well affect the ability of the virus to grow anyway. So, therefore how would you make the virus to release in the field? Having said this, it is possible to control the expression of a toxin gene in the baculovirus and we are working on that now, but those data are not ready to be discussed at the present time.

Intervention:
I would like to make the comment that we sometimes forget that in nature there are a lot of things that are very efficient in killing. For example smallpox was excellent and influenza was far more efficient than all the guns of the first World War but we still survived. The situation is that you have to be able to infect people. And I do sometimes wonder that we get too far carried away about how dangerous something might be and forget the natural problems associated with spread and the ability to infect and to pass on.

CONCLUDING REMARKS ON BACULOVIRUS

Prof. dr. J.E. Beringer (rapporteur), School of Biological Sciences, University of Bristol, United Kingdom

1. Our knowledge of the ecology of baculovirus was not sufficient to make sensible predictions.

2. Baculoviruses and hosts evolved well together. We knew of no evidence that they are able to completely eliminate the host population. Thus we did not expect gene manipulation to alter this significantly.

3. It is possible that baculoviruses can become latent and survive for generations within insect hosts. Insect populations could be infected by them after many generations.

4. There is no way at present that short term risk assessments can guarantee long term risks as the scientific basis is too weak. Long term is beyond one year.

5. With regard to baculovirus, the notion of keystone-species could not be supported.

6. Despite very good molecular knowledge, including complete gene sequence of viral genomes, very little can be predicted about the biology and ecology of viruses which have been modified genetically.

7. The best way to monitor is by means of field recording by concerned persons (e.g. farmers), who are aware of natural fluctuations and cause and effect.

8. The long-term implications of the use of genetically modified baculoviruses will be affected by the extent of their use. This will depend on developments in chemical pesticides and the production of insect-resistant plants.

TOPIC 4

VACCINES

REGULATION OF THE DEVELOPMENT OF ENVIRONMENTALLY SAFE GENETICALLY ENGINEERED VACCINES: THE ACQUIRED EXPERIENCE

Dr. W. Moens (rapporteur), Institute of Hygiene and Epidemiology, Ministry of Public Health and Environment, Belgium

The spreading of products containing genetically engineered viruses in absence of physical containment raises many environmental and health questions.
The integration of appropriate biological, genetical and ecological confinement into the vaccine itself and the targeting is being assessed through the monitoring of deliberate releases in the environment.

Up to now, there is only one mature example of such a vaccine, it is the recombinant anti-rabies vaccines, commercially named "RABORAL".
Historically, the Raboral's recombinant virus was the creation of the late Jean Pierre Lecocq, a Belgian geneticist and co-founder of the company "Transgene" in Strasbourg. The vaccine itself was set up by Transgene in collaboration with Rhone-Mérieux in Strasbourg. The latter has designed the veterinarian bait that has been clinically tested in Belgian forests by professor Pastoret's team on behalf of the Belgian Ministry of Agriculture as the applicant.
Subsequently, the releases organized in Belgium and later in France were to be controlled by the national and federal authorities of Public Health and Environment. In November 1989 and before the final vote of directive 90/220/EEC, the availability of European biosafety criteria for the release of genetically modified organisms into the environment have been applied by the Belgian regulatory authorities to frame all subsequent releases of the vaccine. EC biosafety provisions were found appropriate to cope with the safe development of Raboral.

For the first time in 1993, rabies has been declared potentially eradicated in Belgium. The same year, Rhone-Mérieux has notified Belgium for Raboral's environmental clearance according to the procedure defined under part C of directive 90/220/EEC. Experimentation of Raboral in Europe and in the United States and all after-release monitoring studies have indicated that this product is safe for the environment and public health. Therefore, on September 17, 1993, the regulatory committee of the European competent authorities for directive 90/220/EEC has adopted the proposal of the Commission to authorizing the placing of Raboral on the European market. Between the cloning of the rabies glycoprotein gene into a vaccinia virus and its European clearance, 13 years have passed.

In the mean time, the European Community has established a regulatory framework for the use and the placing on the market of GMOs. The largest EC member states have already implemented and applied this framework in their legal practices. The practical application of directive 90/220/EEC in Belgium has shown that normal implementation of such a directive should not impair any careful vaccine development.

The use of Raboral in Europe is logically expected to stop -or to be dramatically reduced- once rabies eradication will have been reached. However, vaccination using non-environmentally-regulated attenuated rabies virus is still being carried out in large areas of Europe, although its relative safety and efficiency might now legitimately put into question. Therefore, vaccination campaigns are still planned in Belgium, although at a much lower scale, chiefly along the Belgian eastern borders.

The clinical and biosafety experience acquired with the large scale experimentation of Raboral will be documented by the speakers. Professor Pastoret will report on the design and the clinical research regarding Raboral, whereas dr. Kimman will focus on the ecological long-term effects of the application of genetically engineered vaccines.

POTENTIAL LONG TERM ECOLOGICAL IMPACTS OF THE RELEASE OF A RECOMBINANT VACCINIA-RABIES VIRUS FOR WILDLIFE VACCINATION AGAINST RABIES

Prof. dr. P.P. Pastoret, dr. D. Boulanger and dr. B. Brochier, Department of Virology and Immunology, Faculty of Veterinary Medicine, University of Liège, Belgium

INTRODUCTION: RABIES A WORLDWIDE PROBLEM

Rabies is a fearful disease still prevailing in many countries inmost part of the world. Rabies may be maintained in two not necessarily inter-related cycles, urban and sylvatic. Urban rabies, affecting stray and feral dogs and cats, is by far the most dangerous to man, accounting for an estimated 99 per cent of all recorded human cases and for 92 per cent of all human post-exposure treatments. Sylvatic rabies is characterized by the involvement of one or two main wild species in particular locations and this pattern remains stable over many years (King and Turner, 1993). Wild animal species involved in the maintenance of the infection may vary according to geographical and ecological conditions. For instance there exists in Latin America a sylvatic rabies cycle linked to vampire bats, mainly the species *Desmodus rotundus*.

In Northern America, several wildlife species play a distinct role, such as the raccoon (*Procyon lotor*), the striped skunk (*Mephitis mephitis*), the red fox (*Vulpes vulpes*), the coyote (*Canis latrans*) and the Arctic fox (*Alopex lagopus*).

The present European terrestrial epizootic of rabies has spread some 1,400 kilometres westward from Poland since 1939. For several years, the front of the epizootic advanced 20 to 60 kilometres per year (Pastoret et al., 1989). This terrestrial epizootic is sylvatic: the reservoir of infection is in wildlife. While it involves all susceptible species both wild and domestic, the red fox (*Vulpes vulpes*) is involved in more than 75 per cent of cases. The red fox is both the vector of the disease and its reservoir. It plays a key role in the maintenance of the disease but usually it does not transmit it directly to man. Man is mainly at risk from affected domestic animals such as cattle and cats. The fox seems to be the only species acting in the maintenance of the present terrestrial epizootic; if rabies were to be eliminated from the population of foxes it would cease to be a problem in other wildlife or domestic species. The control of fox rabies will be taken as an example; nevertheless, we must keep in mind that, taking into account the fact that many different epidemiological cycles exist in the world, either rural or sylvatic, involving many different animal species, we must try develop control measures (for instance through vaccination) which can be applied in as many different situations as possible.

RABIES VIRUS

Rabies virus belongs to the family *Rhabdoviridae*. *Rhabdoviridae* are enveloped RNA viruses characterized by their shape (Greek "Rhabdos": rod), and by the presence of helical nucleocapsids that are infectious and enclosed in a lipid envelope bearing surface projections. The genome consists of a single molecule of negative single-stranded RNA which is non infectious and is transcribed into five mRNAs, each of which codes for a single protein. The genome of rabies virus which has been completely sequenced contains 11932 nucleotides (Tordo et al., 1988). The presence of a pseudogene between the G and L cistrons implies that rabies virus is an evolutionarily intermediate in the Rhabdoviridae family. The helical core of the ribonucleoprotein (RNP) contains RNA complexed with about 1800 molecules of nucleoprotein N, 30-60 molecules of transcriptase L, and 950 molecules of phosphoprotein NS. The nucleocapsid structure is surrounded by an envelope of about 1500 molecules of membrane protein M. 1800 molecules of the trans-membrane protein, the glycosylated G protein, are spiked through this the envelope (Pastoret and Brochier, 1992).

Pathogenicity of rabies virus is partially related to the G protein since the introduction of a mutation at arginine 333 of this protein (selection of mutants resisting neutralization by appropriate monoclonal antibodies), renders the virus avirulent for mice and other species (Coulon et al., 1983; Tufferau et al., 1989).

Rabies has been known for centuries as a disease of man and of animals and, for many years, rabies virus was thought to be unique. It is now clear that antigenic variation exists within rabies virus strains as shown with monoclonal antibodies, and the existence of several distinct rabies-related viruses is now fully recognized (Bourhy et al., 1993).

Recent developments in molecular biology have led to the increased use of genetic typing as an epidemiologic tool. Application of these techniques to rabies epidemiology is still in its early stages; there is not yet a general agreement about the area of the genome on which to focus nor how much sequence information is necessary. Nevertheless it is clear that there exist biological variants of the virus adapted to different species. For instance, Blacou and coworkers have shown that the rabies virus strain prevailing nowadays in Western Europe is well adapted to the fox and far less to the dog (Blancou, 1985; Blancou et al., 1983); dogs being rare excreters of fox rabies virus.

CONTROL OF FOX RABIES

The prophylactic measures taken in the past like the destruction of foxes to reduce fox population, did not prevent the spread of the epizootic. During recent years, most of the research on the control of fox rabies has concentrated on the development of methods of vaccination of the fox by the oral route (Steck et al., 1982) and this method has already been extensively used in all the contaminated countries belonging to the European Communities. Research has focused on oral vaccination because it is the only means allowing the immunization of a sufficient proportion (75%) of wild foxes, through the distribution of vaccine baits. Therefore the only vaccines that could be used were either attenuated strains of rabies virus or live vector vaccines. Inactivated rabies vaccines are useless when given by the oral route (Brochier et al., 1985).

In 1986, in order to develop a common strategy for the European Union, a co-ordinated trial of oral vaccination of foxes was undertaken in several European countries using the SAD B 19 attenuated strain of rabies virus in order to assess both the efficacy and the feasibility of the method (Pastoret et al., 1987). The results of these campaign confirmed the efficacy of fox vaccination for the control of sylvatic rabies.

However, the use of attenuated rabies virus remains controversial as far as safety and stability are concerned, as these virus strains are still pathogenic for laboratory and wild rodents (Leblois and Flamand, 1988; Wandeler et al., 1982) or wildlife species such as the chacma baboon (*Papio ursinus*) (Bingham et al., 1992) and target species such as the striped skunk (*Mephitis mephitis*) (Rupprecht et al., 1990); moreover, these strains may still be pathogenic to man. Human beings exposed to SAD derived attenuated strains of rabies must be treated with a conventional inactivated rabies vaccine since it elicits good cross-protective immunity (P. Sureau, personal communication). The SAD derived attenuated strains may also be inefficient for some rabies vectors such as the raccoon (*Procyon lotor*) in North-America (Rupprecht et al., 1986). Due to their residual pathogenicity (Pastoret et al., 1985), attenuated strains of rabies virus are no more used for domestic animals vaccination in Western Europe.

As already mentioned, pathogenicity of attenuated rabies virus strains can be abolished by mutating Arg residues at position 333 of the rabies virus glycoprotein and it has led to the development of a new attenuated vaccine strain already used in the field (Leblois et al., 1990). Nevertheless, another inconvenience of attenuated strains of rabies virus is their heat-sensitivity (Languet, personal communication) which reduces their potential efficacy in field conditions. Thus, in order to improve both the safety and stability of the vaccines used for fox vaccination in the field, a recombinant vaccinia virus which expresses the immunizing glycoprotein of rabies virus has been developed and tested in the field for oral vaccination of foxes against rabies (Kieny et al., 1984; Wiktor et al., 1984; Blancou et al., 1986; Pastoret et al., 1992).

DEVELOPMENT OF A VACCINIA-RABIES VECTOR VACCINE FOR ORAL VACCINATION OF WILDLIFE AGAINST RABIES (TARGET SPECIES)

The glycoprotein of rabies virus is the sole viral protein present on the external surface of the viral membrane. It is the only viral antigen capable of eliciting the production of rabies virus-neutralizing antibodies and has been shown to be capable of conferring immunity to rabies. Thus, the rabies virus glycoprotein is an ideal candidate for use in the construction of subunit, marked vaccine.

Nucleotide sequence analysis of the glycoprotein gene reveals an open reading frame of 524 amino acids. This rabies virus glycoprotein gene has been inserted into the thymidine-kinase (TK) gene of vaccinia virus (VV), generating a selectable TK-virus (Kieny et al., 1984; Wiktor et al., 1984) known as VVTGg RAB.

VVTGg RAB has been tested for efficacy and safety in the main target species in Western Europe and North America: fox, raccoon and striped skunk (Blancou et al., 1986; Tolson et al., 1987; Rupprecht et al., 1988). The duration of immunity conferred by VVTGg RAB, a minimum of 12 months in cubs (Brochier et al., 1988) and 18 months in adult animals (Blancou et al., 1989), corresponds to the length of protection required for fox vaccination in the field, due to the high turnover of the fox population. Oral administration being the only route appropriate for fox vaccination, the vaccine had to be given in a form suitable for ingestion. The efficacy of VVTGg RAB (10^8 $TCID_{50}$) contained in a machine-made baiting system has therefore been tested (Brochier et al., 1990 a) and shown to be fairly efficacious. The baiting-sachet system permits an efficient release of the vaccine virus suspension into the fox mouth.

A vaccine virus must not only be efficacious but also safe for the target species. VVTGg RAB was observed to be non pathogenic in the fox whatever the dose of inoculation (10^2 - 10^{10} $TCID_{50}$), or route of administration (Pastoret et al., 1992).

It is also of major importance to preclude epizootiological risks, such as the emergence of asymptomatic carriers of wild rabies virus. This situation could occur in the field by vaccination of naturally infected animals during the incubation period (Thiriart et al., 1985).
The influence of vaccination with VVTGg RAB on the onset of the disease and on the delay before death in foxes previously infected with wild rabies virus, has been investigated. The results show that "early" an "late" death phenomena occur as a consequence of interactions between oral vaccination with VVTGg RAB and rabies infection, but preclude the risk of the emergence of asymptomatic carriers of wild-rabies virus after vaccination (Brochier et al., 1989a).
It is also preferable that a vaccine virus used for oral vaccination of wildlife is not to be horizontally transmitted to unvaccinated animals. No transmission of immunizing amounts of VVTGg RAB was found to occur in adult or young foxes, with the exception of one adult fox bitten by a freshly inoculated other one (Blancou et al., 1986; Brochier et al., 1988). Other safety studies were done as for the development of a conventional vaccine (Pastoret et al., 1992).

STUDIES ON THE TROPISM OF THE RECOMBINANT VACCINIA-RABIES VIRUS IN THE TARGET SPECIES

Experiments were designed to determine the multiplication site in foxes of the recombinant virus as compared with that of the parental strain of the virus, by virus isolation, titration, and indirect immunofluorescence. The polymerase chain reaction (PCR) was also used to detect specific viral DNA in several fox organs. Foxes were fed with 10^8 $TCID_{50}$ of either VVTGg RAB or VV and were euthanized 12, 24, 48, or 96 h after inoculation (Thomas et al., 1990).

Virus Isolation

Small amounts of virus (10^2 to $10^{4.3}$ $TCID_{50}$/ml) were recovered from the tonsils of five of the seven foxes inoculated with VVTGg RAB, during the first 48 h only. VV was detected in the tonsils of the animal after 24 h. Virus was not detected ($<10^{1.5}$ $TCID_{50}$/ml) in any other organ, serum, or faeces of the vaccinated foxes, nor in any organ of the control animal.

Indirect Immunofluorescence

The indirect immunofluorescence test carried out with anti-VV rabbit polyclonal serum confirmed the presence of the virus in the tonsils of four of the five foxes from which VVTGg RAB had been isolated and in the fox from which VV was isolated. The indirect immunofluorescence test carried out with the anti-rabies glycoprotein monoclonal antibody confirmed the presence of virus in the tonsils of three of the five foxes from which the virus was re-isolated. Immunofluorescence was, however, too diffuse to allow precise localization of the virus multiplication in the tonsil.

Polymerase Chain Reaction

Negative controls consisted of DNA from an uninfected fox. Positive controls were prepared with several dilutions of VVTGg RAB or VV in normal fox DNA; the dilutions ranged from approximately one infectious particle per cell to one infectious particle per 10^6 cells. For each virus (VV or VVTGg RAB) 10 µl of the amplified product of the 10^{-2} dilution (corresponding to 1 infectious particle/100 cells) was detected by direct gel analysis. For each virus 30µl of the reaction mixture submitted to two series of 35 PCR cycles were detected by dot blot hybridization, up to and including the 10^{-6} dilution (1 infectious particle/million cells). No detectable amplification was observed in different samples of normal fox genomic DNA. VVTGg RAB was detected in the tonsils, the buccal mucosa, and the soft palate of foxes. No virus was detected from any other organ nor from any of the organs of the control fox. VV was detected in the tonsils and the buccal mucosa only.

Conclusions of the studies on the tropism of VVTGg RAB

Using different techniques, VVTGg RAB or VV were detected during the first 48 h following vaccination by the oral route, but only in tonsils, buccal mucosa, and soft palate. Similar results have been obtained by others in raccoons, using virus isolation (Rupprecht et al., 1988). Results of other unpublished experiments demonstrate that tonsillectomy of the foxes does not affect the protection conferred by vaccination with the VVTGg RAB. Viremia was never observed on days 0, 2, 3, 4, 5, 6, 7, 8, and 14 in inoculated foxes. All these data suggest that orally administered recombinant virus multiplies locally and at a low level. No virus could be detected in salivary glands (parotid and maxillary glands); the risk of transmission by saliva from one animal to another is therefore very low. Furthermore, the fact that VVTGg

RAB multiplies only in restricted sites minimizes the potential risk of recombination with other orthopoxviruses. In our experiments, no difference was observed in the replication sites of VVTGg RAB as compared to the parental strain of VV, demonstrating that recombination does not modify the tropism of VV.
None of these viruses were detected in the brain, suggesting that VVTGg RAB does not multiply in nerve cells. These results were consistent with those of other studies reporting the absence of detectable cytological abnormalities in cerebrospinal fluid from raccoons orally vaccinated with VVTGg RAB (Hanlon et al., 1989).

SAFETY OF THE RECOMBINANT VACCINIA-RABIES VIRUS FOR NON TARGET SPECIES

Field trials with baits have shown that several non-target wildlife species compete with foxes for bait consumption (Brochier et al., 1990b). It must also be taken into account that, within the orthopoxvirus group, vaccinia virus has a wide range of host species. In fact, bait uptake monitoring and tetracycline (biomarker included within the bait) detection controls, performed after vaccination campaigns, proved that mustelids, wildboars (*Sus scrofa*) and domestic carnivores may ingest the vaccine baits.

Moreover, a meaningful proportion of the baits are partly eaten by small mammals (Kalpers et al., 1987; Brochier et al., 1988b). It is therefore important to verify the safety of VVTGg RAB for non-target species (both domestic and wild).

In fact, non-target effects of a vector vaccine oral for immunization of wildlife may be considered as a worst-case scenario.

Several non-target wild species have been chosen for testing in Europe because of their opportunistic feeding behaviour and their presence in the areas where the vaccine must be distributed (Brochier et al., 1989b); among them wild boar, Eurasian badger (*Meles meles*) and several micromammals.
Safety of the vaccine has been tested in Daubenton's bat (*Myotis daubentoni*), wild boar, Eurasian badger, wood mouse (*Apodemus sylvaticus*), yellow-necked mouse (*Apodemus flavicollis*), bank vole (*Clethrionmys glareolus*), common vole (*Microtus arvalis*), field vole (*Microtus agrestis*), water vole (*Arvicola terrestris*), common buzzard (*Buteo buteo*), kestrel (*Falco tinnunculus*), carrion crow (*Corvus corone*), magpie (*Pica pica*) and jay (*Garrulus glandarius*).
Clinical signs of rabies and/or pox-inflicted lesions were never observed in the vaccinated animals during the observation period (28 days minimum after vaccination). Similar experiments have been carried out with wild species from North America (Artois et al., 1990; Desmettre et al., 1990), including meadow vole (*Microtus pennsylvanicus*), woodchuck (*Marmota monax*), grey squirrel (*Sciurus carolinensis*), ring-billed gull (*Larus delawarensis*), red-tailed hawk (*Buteo jamaicensis*), great horned owl (*Bubo virginianus*), coyote (*Canis latrans*), grey fox (*Vulpes cinereo argenteus*), white-tailed deer (*Odocoileus virginianus*), bobcat (*Lynx rufus*) common opossum (*Didelphis virginiana*), river otter (*Lutra canadensis*).
Recent experiments have also shown that the recombinant virus, administered either

by scarification or by the oral route, is also safe for squirrel monkeys (*Saimiri sciureus*) and for chimpanzees (*Pan troglodytes*) (Rupprecht et al., 1992).
For several species, experiments were performed with in contact control animals (including cows) to test for horizontal transmission of VVTGg RAB (Desmettre et al., 1990), with always the same negative results.

STUDIES ON THE POSSIBILITY OF GENE EXCHANGES

The only remaining perceived risk to be investigated was the eventual recombination of the recombinant virus with a wild orthopoxvirus. For such an event to occur, both parental viruses must multiply during the same period of time into the same cells of the same animal. This risk could now be discarded since there are no serological evidences of orthopoxvirus infection in the fox population (Crouch et al., unpublished results). Moreover, after experimental inoculation of cowpox virus to foxes by the oral routes, it only multiplies at low level, during a short period of time in the mouth cavity only (Boulanger et al., submitted for publication). Taking into account those epidemiological and experimental data it is most unlikely that recombination between VVTGg RAB and an other orthopoxvirus could occur in the vaccinated foxes. It is therefore preferable to choose a recombinant virus which has no wild counterpart such as vaccinia virus which, besides a long history of use in uncontrolled conditions, never established in wildlife (Pastoret et Brochier, 1990). A vector virus unknown in wildlife but with a wide host range is, for safety reasons better than another one isolated from a target species such as the raccoonpox virus (Esposito et al., 1988), still prevalent in the wild (Thomas et al., 1975). The fact that vaccinia virus could be used during more than 150 years without any unwanted ecological impact, such as establishment in wildlife, (long experience) is also in favour of its choice.

DELIBERATE RELEASE OF THE VACCINIA-RABIES RECOMBINANT VIRUS FOR ORAL VACCINATION OF FOXES AGAINST RABIES

Taking into account all the available experimental data concerning the safety of the VVTGg RAB for target and non-target species and its efficacy in foxes, limited field trials of fox vaccination with the recombinant virus were authorized first by the Belgian (Pastoret et al., 1988; Pastoret and Brochier, 1991; Pastoret et al., 1993) and then by the French public health authorities.

The Belgian authorizations were preceded by safety assessment taking into account risk versus benefit from the use of the recombinant vaccinia-rabies virus for fox vaccination against rabies. The risk coming from rabies can be clearly identified and measured and could be reduced by the use of a more efficacious vaccine such as the VVTGg RAB (due to its immunogenicity and its stability) than the already available ones. As far as safety was concerned, there are clear and identified risks associated with the use of conventionally attenuated rabies virus strains such as the SAD B19 strain. This risk can be abolished either by the use of the recombinant vaccinia-rabies virus or by the use of rabies virus strains which are modified at the level of arginine 333 of the glycoprotein. With the safety of the VVTGg RAB being confirmed by these small trials, the Belgian authorities agreed for an enlarged open field trial (Brochier et al., 1990b) (435 km^2 area) in the southern part of the country using 15 baits/km^2. Each bait contained a suspension of 10^8 $TCID_{50}$ o of VVTGg RAB (2.2 ml by volume) within a plastic sachet and 150 mg tetracycline as a long-term biomarker of bait uptake. The vaccine was very stable even following natural freezing and thawing cycles.

The VVTGg RAB vaccine has been shown to retain its capacity to immunize for at least 1 month in field conditions, a period which corresponds to the delay of uptake that many baits may undergo in the field. Following this enlarged trial, three fox-vaccination campaigns using VVTGg RAB were then carried out in Belgium in November 1989, April 1990 and October 1990 in order to check for efficacy in an area of 2.200 km^2 (Brochier et al., 1991).

TOWARDS ERADICATION OF RABIES?

The last trial of deliberate release of the VVTGg RAB on a 2200 km^2 of southern Belgium intended to test the feasibility of rabies eradication on a large enough area (Brochier et al., 1991). The 25.000 baits containing VVTGg RAB and tetracycline as a biomarker were dropped by helicopter on three occasions (November 1989, April 1990 and October 1990). After the third phase of vaccination, 81 % (64/79) of inspected foxes were tetracycline positive. Only one rabid fox was recorded, at the periphery of the baited area, which was tetracycline negative.

Despite the dramatic decrease in the number of rabid foxes recorded after vaccine-bait distribution, the efficacy of the vaccination campaign remains difficult to evaluate because systematic collection of foxes is not logistically feasible. Nevertheless, because notification of cases of rabies in cattle and sheep is mandatory in Belgium, the incidence of rabies in livestock provides a reliable indicator of the prevalence of rabies in the wild. No case of livestock rabies has been recorded in the study zone since the second phase of vaccination.

On this occasion, we have also investigated the economics of the vaccine-bait dispersal programme. The average yearly cost of rabies in Belgium (1980-89), including post-exposure treatments of humans, animal diagnosis, compensation to farmers for the culling of infected livestock, and the culling of wild foxes, is estimated to be 400,000 ECUs 10,000 km^{-2}, or 88,000 ECUs per annum for the area under study. These figures do not include the cost of vaccination of domestic animals nor the salaries of civil servants. In comparison we estimate the overall expenditure during the three campaigns of vaccine-bait distribution to be 118,000 ECUs. Because vaccination following eradication can, in principle, be interrupted or subsequently limited to the borders of vaccinated zone, long-term maintenance of a rabies-free area by peripheral vaccination with VVTGg RAB is economically justifiable.

The use of VVTGg RAB has now been extended to the whole contaminated areas in Belgium and Grand-Duchy of Luxembourg as well as on large areas in France. As far as Belgium is concerned (Brochier et al., 1993), it is nearly reaching the stage of elimination, since only 4 cases of rabies were reported during the second semester of 1992 (3 foxes, 1 cattle) and only 1 (a badger) during 1993, in the same area where the last cases of fox rabies were reported in 1992. The eradication of rabies in Belgium has already had other beneficial effects, besides improvement of animal health. First, the number of human post-exposure treatments decreased according to the decrease of rabies incidence in animals (mainly cattle). Second, the diminution of rabies incidence in wildlife had a beneficial effect on the survival of threatened wild species in the contaminated area, such as the European badger. Estimation of the badger's population in the treated area shows a gradual increase in number (Bauduin et al., unpublished results). In fact Belgium is recovering the earlier situation of before 1966 when rabies was reintroduced from Germany.

CONCLUSIONS

Today, much of the research required for the initiation of programmes to control and then eradicate both urban and sylvatic rabies is complete. What is needed now is education to change public awareness of the hazards of the disease. The control methods must be socially and internationally acceptable, legally enforceable and economically feasible. No control programme is likely to succeed without international cooperation.

In Western Europe, rabies is considered a source of economic loss and, above all, hampers the movement of animals between the different member states of the European Union. This has serious implications for the "open market" since some member states are currently rabies-free and wish to maintain their disease-free status. Therefore, the control of rabies requires a common strategy established at the level of European Union. There does not seem to be potential long term ecological impacts involved directly linked to the release of the vaccinia-rabies recombinant virus in wildlife. The impact could be indirect, due to an increase of fox population: regulatory mechanisms other than rabies may therefore play a role in fox population reduction, such as sarcoptic mange (Lindstrom and Morner, 1985). Other indirect effect may be increasing incidence of fox infestation with human pathogens such as *Echinococcosis*, but this is purely speculative and human contaminations can easily be avoided (Brochier et al.; 1992).

REFERENCES

Artois M., et al., Vaccinia recombinant virus expressing the rabies virus glycoprotein: safety and efficacy trials in Canadian wildlife. Can. J. Vet. Res., 1990, 54, 504-507.

Bingham J. et al., The pathogenicity of SAD rabies vaccine in chacma baboons (*Papio ursinus*) given by the oral route. Vet. Rec., 1992, 131, 55-56.

Blancou J et al., La rage du renard. Ann. Med. Vet., 1985, 129, 293-307.

Blancou J. et al., Innocuité et efficacité du virus recombinant vaccine-rage administre par voie orale chez le renard, le chien et le chat. Ann. Rech. Vet., 1989, 20, 195-204.

Blancou J. et al., Différences dans le pouvoir pathogène de souches de virus rabique adaptées au renard ou au chien. Ann. Virol., 1983, b, 134E, 523-531.

Blancou J. et al., Oral vaccination of the fox against rabies using a live recombinant vaccinia virus. Nature, 1986, 322, 373-375.

Boulanger D. et al., Comparisons between the susceptibilities of the red fox (*Vulpes vulpes*) to vaccinia-rabies recombinant virus and cowpox virus. Submitted.

Bourhy H. et al., Molecular diversity of the Lyssavirus Genus. Virology, 1993, 194, 70-81.

Brochier B. et al., Interaction between rabies infection and oral administration of vaccinia-rabies recombinant virus to foxes (*Vulpes vulpes*). J. Gen. Virol., 1989a, 70, 1601-1604.

Brochier B. et al., Use of recombinant vaccinia-rabies glycoprotein virus for oral vaccination of wildlife against rabies: innocuity to several non-target bait consuming species. J. Wild. Dis., 1988b, 25, 540-547.

Brochier B. et al., Programme d'éradication de la rage en Belgique par la vaccination du renard: bilan 1992. Ann. Medd. Vet., 1993, 137, 285-291.

Brochier B. et al., Enquête sur l'infestation du renard roux (*Vulpes vulpes*) par *Echinococcus multilocularis* en province de Luxembourg (Belgique). Ann. Med. Vet., 1992, 136, 497-501.

Brochier B. et al., Vaccination of young foxes (*Vulpes vulpes L.*) against rabies: trials with inactivated vaccine administered by oral and parenteral routes. Ann. Rech. Vet., 1985, 16, 327-333.

Brochier B. et al., Large-scale eradication of rabies using recombinant vaccinia-rabies vaccine. Nature. l99la, 354, 520-522.

Brochier B. et al., Efficacy of a baiting system for fox vaccination against rabies with vaccinia-rabies recombinant virus. Vet. Rec., 1990a, 127, 165-167.

Brochier B. et al., Use of recombinant vaccinia-rabies virus for oral vaccination of fox cubs (*Vulpes vulpes L.*) against rabies. Vet. Microbiol., 1988, 18, 103-108.

Brochier B. et al., Use of vaccinia-rabies recombinant virus for the oral vaccination of foxes against rabies.Vaccine, 1990b, 8, 101-104.

Brochier B. et al., A field trial in Belgium to control fox rabies by oral immunization. Vet. Rec., 1988b, 123, 618-621.

Coulon P. et al., Molecular basis of rabies virulence. II. Identification of a site on the CVS glycoprotein associated with virulence. J. Gen. Virol., 1983, 64, 693-696.

Desmettre P. et al., Use of vaccinia-rabies recombinant for oral vaccination of wildlife. Vet. Microbiol., 1990, 23, 227-236.

Esposito J.J. et al., Successful oral rabies vaccination of raccoons poxvirus recombinants expressing rabies virus glycoprotein. Virology, 1988, 165, 313-316.

Hanlon C.A. et al., Cerebrospinal fluid analysis of rabid and vaccinia-rabies glycoprotein recombinant, orally vaccinated raccoons (*Procyon lotor*). Am. J. Vet. Res., 1989, 50, 364.

Kalpers J. et al., Première campagne de vaccination antirabique du renard par voie orale menée en Belgique. Controles d'innocuité chez les rongeurs et les insectivores. Am. Med. Vet., 1987, 131, 473-478.

Kieny M.P. et al., Expression of rabies virus glycoprotein from a recombinant vaccinia virus. Nature, 1984, 312. 163-166.

King A.A. et al., Rabies: a review.J. Comp. Path., 1993, 108, 1-39.

Le Blois H. et al., Oral immunization of foxes with avirulent rabies virus mutants. Vet. Microbiol., 1990, 23, 259-266.

Le Blois H. et al., Studies on pathogenicity in mice of rabies virus strains used for oral vaccination of foxes in Europe. In: Pastoret P.P., Brochier B., Thomas I., Blancou J. (eds). Vaccination to control rabies in foxes, 1988. Office for Official Publications of the European Communities, Brussels, Luxembourg, pp. 101-104.

Lindstrom E. et al., The spreading of sarcoptic mange among swedish red foxes (*Vulpes vulpes*) in relation to fox population dynamics. Rev. Ecol. (Terre vie), 1985, 40, 211-216.

Pastoret P.P. et al., Rhabdovirus, infection and immunity. In Encyclopedia of Immunology, Edited by Roitt I.M. and Delves P., Academic Press,London, San Diego, New York, Boston, Sydney, Tokyo, Toronto, 1992.

Pastoret P.P. et al., Le virus de la vaccine et ses proches parents. Ann. Med. Vet., 1990, 134, 207-220.

Pastoret P.P. et al., Biological control of wild animal infections. Curr. Opin. Biotechnol., 1991, 2, 465-469.

Pastoret P.P., Development and deliberate release of a vaccinia-rabies recombinant virus for the oral vaccination of foxes against rabies. In. Recombinant Poxviruses, Binns M.M., Smith G.L., Eds, 1992, CRC Press, Boca Raton, Ann Arbor, London,Tokyo.

Pastoret P.P. et al, Deliberate release of recombinant vaccinia-rabies virus for vaccination of wild animals against rabies. Microb. Releases, 1993, 1, 191-195.

Pastoret P.P. et al., First field trial of fox vaccination against rabies using a vaccinia-rabies recombinant virus. Vet. Rec., 1988, 123, 481-483.

Pastoret P.P. et al., Fox rabies in Europe. Irish Veterinary Journal, 1989, 42, 93-95.

Pastoret P.P. et al., Campagne internationale de vaccination antirabique du renard par voie orale menée au Grand-Duché de Luxembourg, en Belgique et en France. Ann. Med. Vet., 1987, 131, 441-447.

Pastoret P.P. et al., Les problèmes associés a la vaccination antirabique des animaux domestiques. Ann. Med. Vet., 1985, 129, 361-374.

Rupprecht C.E. et al., Ineffectiveness and comparative pathogenicity of attenuated rabies virus vaccines for the striped skunk (*Mephitis mephitis*). J. Wildl. Dis., 1990, 20, 99-102.

Rupprecht C.E. et al., Efficacy of a vaccinia-rabies glycoprotein recombinant virus vaccine in raccoons (*Procyon lotor*). Rev. Infect. Dis., 1988, 10, 803-809.

Rupprecht C.E. et al., Primate response to a vaccinia-rabies glycoprotein recombinant virus vaccine. Vaccine, 1992, 10, 368-374.

Rupprecht C.E. et al., Oral immunization and protection of raccoons (*Procyon lotor*) with a vaccinia-rabies glycoprotein recombinant virus vaccine. P.N.A.S., 1986, 83, 7947-7950.

Steck F. et al., Oral immunization of foxes against rabies: laboratory and field studies. Comp. Immun. Microbiol. infect. Dis., 1982, 5, 165-171.

Thiriart C. et al., Immunization of young foxes against rabies: interaction between vaccination and natural infection. Ann. Rech. Vet., 1985, 16, 289-292.

Thomas E.K. et al., Further characterization of raccoon poxvirus. Arch. Virol., 1975, 49, 217-227.

Thomas I. et al., Multiplication site of the vaccinia-rabies glycoprotein recombinant virus administered by the oral route in foxes. J. Gen. Virol., 1990, 71, 3742.

Tolson N.D. et al., Immune response in skunks to a vaccinia virus recombinant expressing the rabies virus glycoprotein. Can. J. Vet. Res., 1987, 51, 363-366.

Tordo N. et al., Molecular genetics of the rabies virus, a century after Pasteur. In: Molecular Biology and Infectious Diseases (ed Schwartz H.), Elsevier, Amsterdam, 1988.

Tuffereau C. et al., Arginine or lysine in position 333 of ERA and CVS glycoprotein is necessary for rabies virulence in adult mice. Virology, 1989, 1 72, 206-212.

Wandeler A. et al., Small mammals studies in a SAD baiting area. Comp. Immunol. Microbiol. infect. Dis., 1982, 5, 173-176.

Wiktor T.J. et al., Protection from rabies by a vaccinia virus recombinant containing the rabies virus glycoprotein gene. P.N.A.S., 1984, 81, 7194-7198.

POTENTIAL LONG TERM ECOLOGICAL EFFECTS OF THE APPLICATION OF GENETICALLY ENGINEERED VACCINES

Dr. T.G. Kimman, Central Veterinary Institute, Department of Virology, The Netherlands

SUMMARY

Genetically engineered vaccines are used to prevent or to minimize the chance on infection or, should infection still occur, to minimize the consequences of infection. In most cases they will be used in confinement under more or less well controlled conditions, such as stables. In some cases, such as with the vaccinia-rabies recombinant, they are released in the environment. Because pathogenic microorganisms are part of the ecosystem, they are thus released with the deliberate purpose to influence the biodiversity of an ecosystem. The vaccinia-rabies recombinant has been applied on a large scale in western Europe with the purpose to eradicate rabies virus among foxes (*Vulpes vulpes*). Before genetically engineered vaccines are released, it is of utmost importance to perform a risk assessment following a "worst case scenario". Short, mid and long term consequences which should be evaluated and minimized, are pathogenic effects of the vaccine strain in the vaccinated species and in species which may unintentionally take up the vaccine, transmission of the vaccine strain in these species, and gene exchange with related and unrelated wild type microorganisms. Such a risk assessment should minimize the chance that a plague occurs in the vaccinated or other species. Establishment of the vaccinia-rabies recombinant does not appear to have happened and it appears unlikely that genetic material from this recombinant will be frequently transferred. Because very unlikely events may still occur at some time, it is nonetheless advisable to stop vaccination as soon as the preset goal has been achieved. Methods are available and will be discussed to reduce risks, including the use of non-transmissible vaccines, a parenteral route of application, and methods to reduce the chances that viable recombinants arise. Other long term consequences may be the result of the disappearance of the pathogen and are more difficult to evaluate. If the pathogen, against which the vaccine is applied, plays an important and unique regulatory role in determining the size or fitness of the target population, then vaccination may have important sequelae leading to temporal ecological imbalances. It is doubtful whether rabies virus has such a role in foxes. The potential long term drawbacks of rabies vaccination of foxes, using a genetically engineered vaccine or not, may therefore be small and should be compared with the advantages of the disappearance of rabies virus.

INTRODUCTION

Human welfare and animal production and welfare are severely hampered by infectious diseases, both in developed and underdeveloped countries. Vaccines are among the most effective and cheap prophylactic tools to help preventing infectious diseases. Vaccines are not only applied to control clinical signs, but they are also applied to reduce transmission of a pathogen in its host species with the purpose to eliminate the pathogen. Examples of successful disease eradication using vaccines are smallpox among humans and, more recently, rabies among foxes in several parts of Europe (Pastoret & Brochier, 1991). Attempts to eradicate Aujeszky's disease among pigs in commercial pig farms using vaccination have also given promising results.

To develop a live vaccine, attenuation is required of the virulent microorganism to which immunity must be induced. Spontaneous attenuation may occur during cultivation of the microorganism and is due to genetic changes varying from point mutations to the loss of genes. Attenuation can also be achieved by deletion mutagenesis using rec DNA techniques. In principle, such genetically engineered deletion mutants do not carry other or more risks than conventionally attenuated mutants. In contrast, rec DNA techniques may be able to construct mutants which will less likely cause safety problems. Well chosen, multiple deletions in several virulence genes may lead to vaccine strains that will very unlikely revert to virulence, whereas conventionally attenuated vaccine strains may have genetic changes that are insufficient to cause attenuation or that are so small that they may easily revert to virulence. For example, the conventional attenuated live Sabin vaccine strain against poliovirus contained only a single nucleotide substitution and could therefore easily revert to virulence (Evans et al., 1985). Reversion to virulence has also occurred in rabies vaccines (Report WHO meeting 1990). Another potential drawback of conventionally attenuated vaccine strains is insufficient attenuation. An early measles vaccine still caused fever and rash in vaccinated children and vaccinia virus caused encephalitis in a small minority of vaccinated people (Mims, 1986).

Vaccination with live carrier vaccines which express genes of foreign pathogens is a promising approach. Complex antigens are amplified in vivo and in a way that mimics antigen presentation after natural infection. This approach may be valuable for the development of vaccines against pathogens which cannot be grown easily or safely in vitro. Most attention has been focused on vaccinia virus, a poxvirus, as vaccine vector. Vaccinia virus is effective in eliciting both B and T cell-mediated responses. Other vectors currently under investigation are adenoviruses, herpesviruses, and several bacteria including *Salmonella*, *Shigella* and *Escherichia coli* species (Report WHO meeting, 1990). Recombinant vaccinia virus carrying the glycoprotein G gene of rabies virus induces complete and long-term resistance against rabies in foxes and raccoons. Recently, vaccine baits containing capsules with recombinant virus have been used successfully in Belgium, Luxembourg, and France to immunize foxes. The incidence of rabies declined rapidly to almost zero (Pastoret & Brochier 1991).

Live carrier vaccines carry similar potential risks of harmful side effects as live non-carrier vaccines, but the incorporation of foreign genes leads to additional safety considerations. Through the incorporation of foreign genes the carrier organism may obtain altered biological properties. The following potential non-target effects of carrier vaccines should be examined:

1. Changes in cell, tissue, or host tropism, and virulence of the carrier organism through the incorporation of foreign genes.

2. Recombination with related microorganisms leading to gene transfer and changes in biological properties of the parent microorganisms.
3. Spread and survival.

Especially survival, spread and recombination may lead to potential long term ecological effects. It is clear the chances on long term effects by this route are very small when the vaccine organism has poor chances for survival. However, also the disappearance of the pathogen or the reduced damage caused by it may lead to changes in an ecosystem.

POTENTIAL LONG TERM EFFECTS OF GENETICALLY RECOMBINANT VACCINES

Non-target effects of recombinant vaccines

Only few experimental data are available which illustrate that the biological behaviour of a microorganism can alter through the incorporation of a foreign gene. For example, there are some indications that vaccinia virus carrying the attachment protein G of respiratory syncytial virus, a pneumotropic virus, replicates better in lungs of mice than vaccinia virus carrying other genes of this virus (Taylor et al., 1991). Likewise, incorporation of the gE virulence gene of pseudorabies virus (PRV) in vaccinia virus appeared to enhance its virulence (Kost et al., 1989).

It is unfortunately seldom known which properties determine cell and tissue tropism and pathogenicity of a microorganism. Current knowledge, therefore, seldom if ever allows to predict a possible change in biologic behaviour of a carrier organism through the incorporation of foreign genes. However, such changes may in theory predominantly be expected when genes encoding surface glycoproteins involved in virus-cell interactions or host range genes are incorporated, although potential differences in processing or cleavability of such proteins in different host cells precludes any prediction. In addition, cell tropism is not only determined by the interaction of an attachment protein with a cellular receptor. Especially expression of the foreign protein on the surface of the vector organism may lead to changes in tropism. Although some purified preparations of recombinant vaccinia virus expressing rabies glycoprotein G have been found to contain the rabies glycoprotein in their envelope, the neurovirulence of the carrier was not altered (Kotwal et al., 1989). In many species the vaccinia-rabies recombinant proved avirulent. This result was not unexpected since the insertion of the gene encoding the rabies virus glycoprotein rendered the vaccinia virus thymidine kinase (TK) negative, which further attenuates the carrier vaccinia virus.

Recombination and reassortment between viruses requires that the same cell is infected by both parent strains. The frequency of recombination between two virus strains is difficult to assess and predict, but will be determined by the virus family, the extent and site of multiplication, the occurrence of persistence and latency, and the routes of infection. In pox- and herpesviruses there appears to be a strong connection between the level of DNA replication and recombination. In nature, new variant influenza strains may emerge from reassortment. Pigs and poultry are the potential reservoir of influenza viruses, from which by reassortment new variant strains may emerge that have an enhanced virulence for humans or mice. The process of reassortment is used for the manufacture of influenza virus vaccines for use in humans. Isolates from camels in Mongolia proved to be derived from the vaccine strain used to manufacture killed virus for use in humans (C. Scholtissek, personal communication). Influenza vaccination has been shown to reduce the replication of influenza virus and may thus reduce the chance of reassortment events between influenza viruses. Exchange of heterologous genetic information between an attenuated carrier microorganism and its wild-type variant may lead to a virulent "wild-type" mutant carrying a foreign gene and potentially expressing altered biological properties. Moreover, such a "worst case" mutant could be transmitted as the wild-type strain and it could even have a competitive advantage over the wild-type strain. Such a competition advantage is very unlikely if the foreign gene is expressed in genes that have a function in replication and virulence, such as the TK gene.

To estimate the risks of recombination it is necessary to know the reservoirs of potential acceptors of foreign genes. For example, vaccinia virus might easily recombine with cowpox virus which is endemic in Europe in small rodents (bank voles, field voles, woodmice) and which may infect domestic cats and man. In fact, cattle are only rarely infected with cowpox virus (A. Crough, personal communication). After oral administration limited replication of the vaccinia-rabies recombinant was demonstrated in tonsils, buccal mucosa and soft palate of foxes. Also cowpox virus exhibited a very limited replication in foxes at restricted sites. The limited replication of these strains suggests that the risk on recombination of the vaccinia-rabies recombinant with other orthopoxviruses is small (D. Boulanger, personal communication). Should transfer of the heterologous gene insert to cowpox virus occur, then the resulting strain is expectedly TK negative, which probably strongly reduces replication and transmissibility of this strain. Moreover, a serological survey indicated that the prevalence of orthopoxvirus in foxes is very low or absent.

Transmission of a microorganism is not only determined by the organism itself and its host range, but also by the dose, the route of application and the level of immunity of the vaccinated individual and population. Mathematical models may predict the spread of a microorganism from small scale laboratory experiments. The wide host range of vaccinia virus has been suggested as a possible cause of problems, i.e. transmission from target animals to non-target species potentially followed by recombination. Vaccinia virus presently exists in Europe only in the laboratory, indicating that is has not established itself after its use as vaccine in humans. However, there is some evidence that the virus may have established itself in the buffalo in India, from which occasional human infection may occur (Baxby, et al., 1986). The vaccinia-rabies recombinant did not detectably spread from vaccinated to non-vaccinated foxes, dogs, cattle, ferrets, wild boars, and badgers in laboratory experiments (Pastoret et al., 1992).

Effects due to pathogen eradication

Most bacteria and viruses living in the environment of man and animals do not cause harm. This can be expected because a successful microbe must try to maintain itself in nature and therewith its host on which it lives. Thus, microbes try to reach a state of balanced pathogenicity in the host, and cause the smallest amount of damage compatible with the need to maintain itself. Simultaneously, the host requires greater resistance against the microbe. A famous example is the case of myxomatosis in the Australian rabbit. Cholera in man appears to have evolved similarly. Disease is considered as an event in evolution in which the infection not yet had the time to reach the ideal state of balanced pathogenicity. Spread of a microorganism into a new continent has sometimes resulted in infection of a new population in whom disease is much more severe because of greater genetic susceptibility. Examples include tuberculosis spreading from resistant Europeans to susceptible Africans and the infection of Northsea seals by phocid distemper virus. Severe disease may occur in species which are irrelevant for the survival of the microorganism, such as rabies or Lassa fever in man, or when a microbe is adapting itself to a new species, for example new influenza virus strains in man or Lelystad virus in swine. It is clear that microbes continue to have a rapid rate of evolution.

Vaccines may be a major antimicrobial force. Several pathogens have been eradicated or are being eradicated by vaccines, examples being smallpox and poliovirus in man and pseudorabies virus in swine. Therewith vaccines will have a major impact on the health of the vaccinated population. This may have unpredictable sequelae, especially if the pathogen is a factor which limits the size of the host population. Eradication of a pathogen may have several sequelae including growth of the host population, increased competition for food, territorium, and sexual partners, and greater chances for predators and other pathogens, newly evolving or not. However, it is clear that eradication of the pathogen is often desirable. The sequelae of pathogen eradication are not or only very poorly predictable. During rabies vaccination of foxes in Belgium, the fox population has reportedly increased (P.P. Pastoret, personal communication). However, it is doubtful whether rabies virus is a major factor controlling the size of the fox population.

Another potential effect of vaccination may be an evolutionary force towards strains with a higher capacity for transmission and survival despite the herd immunity afforded by vaccination. This might more readily occur with RNA than with DNA viruses.

The possibility of influencing population size can also be taken as an advantage. For example, the eradication of rabies from the African wild dog (*Lyacon pictus*) in certain East African game reserves may save this species from extinction (Anderson, 1991).

MINIMIZING THE RISK OF RECOMBINANT VACCINES

Thorough knowledge of the function of the inserted gene product and of the gene at the insertion locus may minimize non-target effects. For example, genes that may alter host tropism of a pathogen should in generally not be incorporated in a carrier organism, or in such a way that it is harmless, for instance because it is not expressed on the cell or virion surface or because a minimum of protective sequences are expressed either as peptides or in non-functional hybrid proteins.

Carrier organisms may be modified so that their environmental spread is minimal or absent. For example, PRV mutants with a deletion of the essential gene encoding gD may infect cells, but infectious viral particles are not produced (Peeters et al., 1993). Likewise, fowlpox virus does not give rise to a productive infection in non-avian species, but it induces immunity to inserted gene products (Taylor et al., 1988). Highly attenuated strains of vaccinia virus have been developed which are characterized by a lack or very low level of excretion (Sutter & Moss, 1992; Tartaglia et al., 1992). For example, the NYVAC vaccinia strain has been derived from the Copenhagen vaccine strain by deleting 18 open reading frames. Although NYVAC has a dramatically reduced ability to replicate on a variety of mammalian cells *in vitro* and *in vivo*, it has retained the capacity to elicit strong immune responses, both humoral and cellular, to extrinsic antigens (Tartaglia et al., 1992).

Deletions in genes that determine latency, or reactivation, or both, may reduce the period during which the carrier is present in the environment and therewith reduce the risks on recombination. The choice for a vector with a small host range may further limit the environmental spread of the vector, but precludes its general use. For example, swinepox virus only infects swine and would therefore be a suitable carrier candidate in swine. In contrast, vaccinia virus has a very broad host range.

Chances on exchange of genes can be further minimized by choosing a parenteral route of immunization. Should there be a chance on recombination, then the possible harmful sequelae can be minimized. For example, expression of a foreign gene in a gene that determines virulence or replication of a microorganism will yield a less virulent microorganism after recombination with a wild-type microorganism. Finally, carrier organisms may be designed so that lethal mutations occur after recombination with wild-type strains. For that purpose, foreign genes should be inserted in vital genes so that non-viable mutants occur upon recombination.

DISCUSSION AND CONCLUSIONS

1. Recombinant DNA technology has given the possibility to construct mutant microorganisms with well characterized deletions in virulence genes and vector microorganisms that express proteins of other pathogens.

2. The use of deletion mutants with well characterized deletions do not carry other or more risks than the use of conventionally attenuated strains. Because the mutations are known, the likelihood of reversion to virulence

can even be better estimated and predicted. Through the use of large and multiple deletions the likelihood of reversion to virulence will be very small.

3. Through the incorporation of foreign genes a vector organism may in theory acquire an altered host or tissue tropism. Other concerns regarding live vaccines are transmission to non-target animals and possible recombination events, which may in theory lead to spreading of new microorganisms and long term ecological effects. The example of this case study, the vaccinia-rabies recombinant, has been shown to be innocuous for the target and several non-target species.

4. Recombination between viruses requires that the same cell is infected with both parent strains, which will be a very unlikely event in practice, especially when poor replicating vaccine strains are used. Recombination between a vaccine strain and a wild-type strain will unlikely result in more virulent strain than the wild-type strain itself. Moreover, a recombinant strain will have poor chances of survival due to competition with the wild-type strain. Chances for survival of recombinants are further diminished when a vector with a small host range is used and when the population is completely immunized. Because very unlikely events may occur at some time, it is nonetheless advisable to stop vaccination as soon as the present goal has been achieved.

5. Long-term ecological effects may follow the disappearance of a pathogen from a population, especially if the pathogen limits the size or fitness of that population. Potential effects may include increased population size, leading to lack of space, food, and sexual partners, and increased suscepti-bility for predators or other pathogens. Due to multiple species interaction prediction of these effects is less feasible, if at all.

6. Risks connected with the use of vector vaccines can be further minimized by the use of highly attenuated strains that do not produce infectious particles, that only partly replicate, or that have an extremely restricted host range. In this way, safer vaccines can be produced than possible by conven-tional methods.

7. The potential risks of vaccination should be compared with the benefits.

REFERENCES

Anderson, R.M. (1991): Immunization in the field. Nature, 354:502-503.

Baxby, D., R.M. Gaskell, C.J. Gaskell, M. Bennet. (1986): Ecology of orthopoxviruses and use of recombinant vaccinia vaccines. Lancet, 11:850-851.

Evans, D.M.A., G. Dunn, P.D. Minor, G.C. Schild, A.J. Cann, G. Stanway, J.W. Almond, K. Currey & J.V. Maizel. (1985): Increased neurovirulence associated with a single nucleotide change in a non-coding region of the Sabin type 3 poliovaccine genome. Nature, 314:548-550.

Kost, T.A., E.V. Jones, K.M. Smith, A.P. Reed, A.L. Brown & T.J. Miller. (1989): Biological evaluation of glycoproteins mapping to two distinct mRNAs within the BamHI fragment 7 of pseudorabies virus: Expression of the coding regions by vaccinia virus. Virology, 171:365-376.

Kotwal, G.J, R.M.L. Buller, A.Z. Kapikian, B. Moss, E. Stephens, R.W. Compans. (1989): Analysis of recombinant vaccinia virions for envelope-associated foreign proteins. In: Modern approaches to new vaccines including prevention of AIDS. Editors: R.A. Lerner, H. Ginsberg, R.M. Chanock & F. Brown. Cold Spring Harbor Laboratory, New York, pp 389-392.

Mims, C.A. (1986): The pathogenesis of infectious diseases. Third edition, Academic press.

Pastoret, P.-P. & B. Brochier. (1991): Biological control of wild animal infections. Current Opinion in Biotechnology, 2:465-469.

Pastoret, P.-P., B. Brochier, J. Blancou, M. Artois, M. Aubert, M.-P. Kieny, J.-P. Lecocq, B. Languet, G. Chappuis, P. Desmettre. (1992): Development and deliberate release of a vaccinia-rabies recombinant virus for the oral vaccination of foxes against rabies. In: Recombinant poxviruses, Editors: M.M. Binns, & G.L. Smith, CRC Press, Boca Raton, Florida.

Peeters, B., A. Bouma, T. de Bruin, R. Moormann, A.L.J. Gielkens, T.G. Kimman. (1993): Non-transmissible pseudorabies virus gp50 mutants: a new generation of safe live vaccines. Vaccine, in press.

Report WHO meeting. (1990): Potential use of live viral and bacterial vectors for vaccines. Vaccine, 8:425-437.

Sutter, G., B. Moss. (1992): Nonreplicating vaccinia vector efficiently expresses recombinant genes. Proceedings National Academy of Sciences USA, 89:10847-10851.

Tartaglia, J., M.E. Perkus, J. Taylor, E.K. Norton, J.-C. Audonnet, W.I. Cox, S.W. Davis, J. v.d. Hoeven, B. Meignier, M. Riviere, B. Languet, E. Paoletti. (1992): NYVAC: A highly attenuated strain of vaccinia virus. Virology, 188:217-232.

Taylor, G., E.J. Stott, G. Wertz & A. Ball. (1991): Comparison of the virulence of wild-type thymidine kinase (tk)-deficient and tk+ phenotypes of vaccinia recombinants after intranasal inoculation of mice. Journal of General Virology, 72:125-130.

Taylor J., R. Weinberg, B. Languet, P. Desmettre & E. Paoletti. (1988): Recombinant fowlpox virus inducing protective immunity in non-avian species. Vaccine, 6:497-503.

Recommended literature for further reading

Kimman, T.G. (1992): Risks connected with the use of conventional and genetically engineered vaccines. Veterinary Quarterly, 14:110-118.

Non-Target Effects of Recombinant Live Vaccines. Proceedings Workshop, Langen, 3-5 November 1993, in press.

PLENARY DISCUSSION AT THE END OF THE SESSION

Question:
The result of rabies eradication in Belgium will be the following: other and much more important infections will increase. However, this is not a problem of GMOs but it is a problem of vaccination. Foxes will get seriously infected by a certain worm (of which I don't know the English term) and this worm can also infect human beings. The consequence of this infection may be far more severe than infection by rabies. It is predicted that this worm will cause more human beings to dy in one year than in the case of rabies. The problem is the long incubation time of more than ten to twenty years. So, in Germany we have a discussion that some ecologists are warning against the vaccination of foxes because of the serious, unwanted side effects. Therefore, some people are suggesting to stop fox vaccination campaigns in Germany. One could propose to continue these vaccination provided one similarly develops a cure against this worm but in my view that would be an end-of-pipe solution.

Response:
As I mentioned during my presentation, we are dealing with two different things here. The first thing is the long term effect of the GMO and I think that there are no long term effects associated with the GMO. But there is this other problem: What is the long term effect of the eradication of a disease? For instance, in the case of human beings: Did we do a good thing by eradicating small pox? Or should we live with small pox? The reason why we do vaccinate foxes, I would say, is a political one. The reason why you want to eradicate small pox or to live with it is also political. But of course the long term effects of the eradication of pathogen will have an impact on the population. As far as this worm concerns, we think that we can deal with this problem quite easily. First, it is not a pathogen for foxes, it is only pathogenic to other species like micromammals or man. You can take quite simple precautionary measures to limit the risk of infestation. So far we don't have evidence that we are increasing the level of infestations in foxes. Of course we see an increase of the fox population. I would like to remind you of other beneficial effects. For instance, in Belgium the badger population was decreasing because of hunting and rabies. Now, it is clear cut that their situation is really improving.

Question:
In discussing how to make vaccines safe, the question of recombination has been raised and discussed extensively. But one molecular situation has not been mentioned and that is of 'complementation' of functions between the defective virus and the superinfecting virus. Could the speakers comment upon this?

Response:
Complementation can indeed be demonstrated and the example I have in mind concerns two pseudo rabies strains; both deleted in some genes but not in the same genes. Upon inoculation into one animal it can be shown that both strains can complement each other so that you have a replication of one of the deletion mutants

thanks to the complete viral genome which is present in that particular animal. Although these effects can occur and can be demonstrated, I think the major point is whether it will do any harm and whether either of these strains, the parent strains or the recombinant strain, can be transmitted into the population in which they are introduced and then I think it is very likely that the deletion mutants will have a competition disadvantage compared to the wildtype virus present in that population. In addition, thanks to vaccination or to the wildtype virus present in that population, there is partial or complete immunity against virus transmission in that population.

Question:
Is there any evidence for complementation between less related or even unrelated viruses?

Response:
I guess in the laboratory on the cell or molecular level but as far as I am aware not on the level of the whole animal.

Question:
Could I ask what is the policy in Belgium over the next two years? Will the vaccine continued to be used?

Response:
The policy has not yet been decided. But in some areas we don't vaccinate anymore since more than one and a half year without any reintroduction of rabies. Therefore, we are contemplating to stop the vaccination. Our last vaccination campaign was during this autumn and maybe a next one during spring and then perhaps we just wait. There is no reason why you should go on. As a long term effect of vaccination you can also take into account economic effects because it is an investment. Let me explain: If you live with rabies, a certain percentage of foxes dies from rabies each year; rabies costs about 50 million Belgian francs each year, the campaigns cost around 15 million Belgian frances each year. So, if we manage to get rid of rabies, we will no longer have these costs of the vaccination campaigns, nor of human exposure and treatment and of dying cattle due to rabies contamination and so forth. Together with our European partners France, Germany and Luxembourg we have an agreement to inform each other if you have a case of rabies within 30 kilometres from the border. Therefore, I think we will maintain vaccination in a area of 30 kilometres from the last case.
In relation to ecological concerns, I would like to mention the following: If a disease occurs or comes back, you see it. So, if there will be a reintroduction of rabies in the fox population, we will immediately notice it. Foxes are like Petri dishes for rabies.

Response:
The Competent Authority from the Agricultural Ministry would like to stop the use of the recombinant vaccine as soon as a stable situation of eradication of rabies can be certified. However, if Germany continues to use the attenuated strain to vaccinate the German foxes, we have to protect our borders against infection by these German foxes.

Question:
I though it was very interesting that one of the effects which professor Pastoret mentioned was not so much a long term ecological effect but the indirect effect on other species. I could give some interesting liability questions as to how far you trace the train of consequence. For example, in the United Kingdom there has been a lot of gassing of badgers because they are seen as a reservoir for bovine tuberculoses. I don't know whether you have that as problem of the cattle in Belgium. Making a more general point, that as a result of enhancing the health and the size of the fox population and other populations that were being infected by rabies, namely the badgers, you might have indirect effects with other consequences, economic as well as sanitary.

Response:
In Belgium we don't have problems with bovine tuberculosis associated to the badgers until now, as far as we know. I know the problem in Great Britain. Maybe it will be a new problem: Do you want to live with badgers or not? I have to confess I am in favour of living together with badgers. But it is indeed known that under certain ecological conditions badgers may bring tuberculosis to cattle.

Question:
You showed a slide indicating correctly what a low probability of recombination there was between wild orthopox virus and the recombinant virus because a number of conditions have to be fulfilled at the same time. Perhaps you could elaborate on what the consequence would be if that would happen. So, what does it mean, when that very low probability were to occur? My second question was prompted by something on dr. Kimman's slide. He showed us a statement that a new variant may emerge with the capacity to survive despite the vaccination. Perhaps he might tell us in the context of this particular example whether he has looked at that aspect and what it means. The third point is that it was interesting to see what a large percentage of wild boars seem to have taken up the bait and it occurred to me that they are in fact eaten by humans unlike foxes. So, I wondered whether there is a food aspect to this?

Response:
In fact, I am confident that recombinant vaccinia-rabies virus behaves in foxes like other strains behave. You know that there are new developments in vector viruses; for instance, avipox viruses with an abortive multiplication cycle in the species. I am fully convinced that the same is true for recombinant vaccinia-rabies virus in foxes. This can explain that you have a very small amount of virus after one or two days; it is merely the remaining of the inoculum: 10^2 - 10^4. The second point is that we have the following experimental evidence: You will see that you will protect 100 % of the foxes at 10^7-10^8, only 50 % at 10^6 and only 25 % at 10^5 and no more if you decrease the titer. That is also in favour of an abortive multiplication. Normally an attenuated virus will amplify its antigen due to its own multiplication. And the fact that you have strictly dose-dependent effects is in favour of an abortive multiplication. Therefore, your virus might undergo only one multiplication cycle in some cells of the foxes. So, the risk of recombination will be extremely low, if it

happens. In fact we could show by the same experiment that you don't have a modification of tropism and that the glycoprotein of the rabies virus is not exposed at the membrane level. If you try to have recombination in vitro between recombinant vaccinia-rabies virus and cowpox virus, of course you will succeed. It has already been demonstrated in 1958 that recombination is possible within the orthopoxvirus family with very crude methods. So, if you have a recombinant without modifying the tropism, it will be a dead end phenomenon. Therefore, you give more attenuation to your recombinant than the previous wildtype virus which could recombine with your recombinant. That is a very important point because some people are trying to favour viruses coming from the same species. For instance, in the United States they isolated a racoon poxvirus and they inserted the glycoprotein gene from rabies virus. Then they told the people that it would be more efficacious in racoon, if you use racoon poxvirus but I would say no. It is better to use a vector without a counterpart in the wild, if possible, because there is a greater risk of recombination if the counterpart is still around in the wildlife.

CONCLUSIONS OF THE DISCUSSIONS ON VACCINES

Dr. W. Moens (rapporteur), Institute of Hygiene and Epidemiology, Ministry of Public Health and Environment, Belgium

CONTEXT OF THE DESIGN AND DEVELOPMENT OF RECOMBINANT LIFE VACCINES

Human beings will have serious long-term ecological effects anyhow. Human beings do want to use vaccines to limit the consequences of serious diseases. In doing so, we should try to avoid negative consequences as much as possible.
The development of genetically engineered vaccines has occurred in the context of the one century-old scientific, clinical and commercial context of classical vaccines. The participants to the discussions observe that very similar, if not identical vaccines can be designed using either the classical or modern technologies. Moreover, the development of classical and genetically engineered vaccines require an identical core of clinical science. It was also observed that classical vaccines are not presently submitted to environmental regulations.

IS THERE ANY SPECIFIC ENVIRONMENTAL AND/OR HEALTH PROBLEM RELATED TO THE USE OF RECOMBINANT DNA TECHNOLOGY FOR THE DESIGN OF VACCINES?

For identical vaccines designed by classical or recombinant DNA technology, the participants unanimously stated that there is not such a problem. As an example, the anti-Aujeszky disease vaccine bearing two complementary point mutations could be obtained either by classical selection or by site directed-mutagenesis. In addition, the recombinant vaccines are perceived as safer than classical vaccines. The case of anti-rabies recombinant vaccine was described during the meeting as an environmentally-safer vaccine than its classical counterpart. In a legal framework of Good Manufacturing Practice (GMP), the recombinant vaccines were considered as providing better quality control management of the vaccine production and thus leading to safer products.

FOR WHICH REASONS ARE RECOMBINANT VACCINES PERCEIVED AS SAFER?

Recombinant life vaccines are currently built in micro-organisms like viruses or bacteria. The genetically modified micro-organisms have very specific familiarity and predictability properties in the general context of genetically modified organisms.

The risk associated with the transfer of a genetic information (an average of 300 to 3,000 base pairs) in an acceptor (carrier) organism has to be assessed not only from the nature of the transferred information but also in the context of the acceptor genome, biology and evolutionary history. In recombinant vaccines constructed from viruses, the acceptor (carrier) genome is extremely small when compared to a mouse or a sugar beet genome. Moreover, the ability to design a functional and efficient recombinant organism is proportional to the knowledge of the acceptor (carrier) genome and biology. The natural biodiversity and the plasticity are much greater in the case of micro-organisms than animals and plants. As a consequence, the genetic fluxes associated with microbial ecology, and, consequently, the taxonomic identification of microbes is still a matter of advanced research (EC BRIDGE/BIOTECH Research programs). However, the diseases associated with microbes of vaccine interest have been studied during decades, and at least for these micro-organisms, the familiarity with the acceptor (carrier) ecology and taxonomy is generally very high. As a result of the previous considerations, the predictability of recombinant vaccine in the target and in the environment is relatively very high.

The following scheme summarizes these parameters:

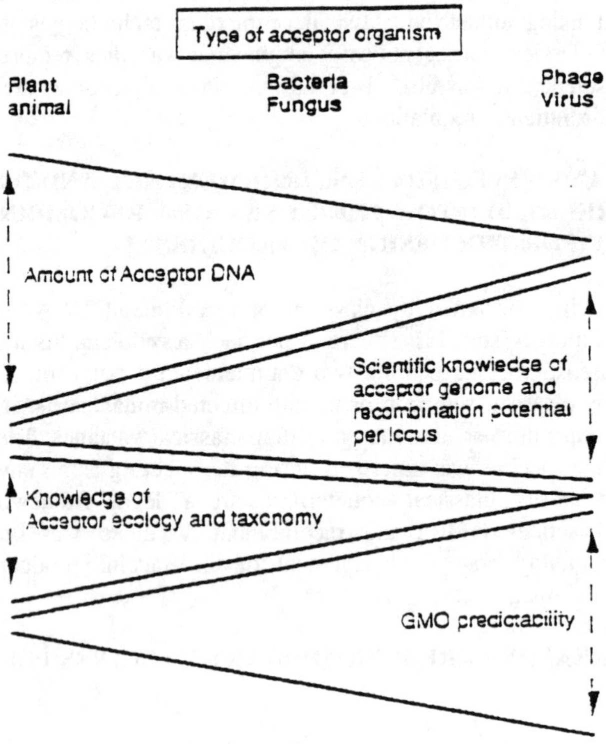

The participants to the discussion consider that safety essentially results from the knowledge of the recombinant vaccine components: inserted or altered gene structure and function, design of the carrier whose safety increases with size reduction, ability to select the carrier, ability to select the carrier as a function of the end target and predictable ecosystems, ability to control the genes responsible for survival, host specificity and recombination, ability to increase the efficiency of the genetically modified micro-organism.

DO WE HAVE A METHODOLOGY TO ASSESS SHORT-TERM AND LONG-TERM IMPACTS?

There is no essential difference between short-term and long-term impacts in the case of recombinant live vaccines. The methodology is available and is similar to the one applied to classical vaccines. However, the participants listed several additional considerations in the case of vaccines that could not have been designed by classical means. The anti-rabies vaccine analyzed during the meeting is such a case.

Tropism

The control of potential changes in the tropism, i.e. accidental or deterministic manipulations of the gene sequences coding for the carrier tropism, interaction of the inserted gene or its product with the gene(s) or function(s) of a receptor-like function.

Recombination potential

The assessment and control of the recombination potential of the recombinant micro-organism: the virus can be selected according to the stability of its taxonomic group or its lack of evolutionary or ecological relationships with the target and target's microbial ecology.

Transmission

The recombinant organism should be selected and/or designed in such a way that it is not transmissible outside the target's body and/or remains completely target-dependent.

CONSIDERATIONS ABOUT SHORT-TERM AND LONG-TERM ECOLOGICAL EFFECTS

In order to assess potential short-term and long-term effects, the properties of inserted gene(s), the carrier, the targeting techniques of the vaccine have to be carefully analyzed.

Target-dependency

The impact of one or another ecological parameter of a vaccine depends on the aim of the vaccine itself. The vaccine can be either used for the eradication of a disease in the wild (the target is the reservoir species of the disease, for example, the anti-rabies vaccine) or for the protection of the injured population (the target differs from the reservoir species). According to the aim of the vaccine, the long-term effects are perceived as different but in both cases as a mixture of predictable and unpredictable effects. But such considerations have nothing to do with the fact that the vaccine was produced by genetic engineering. Genetic engineering was considered to be the fastest, safest, and most economical way to manage the environmentally safe targeting of the vaccine.

Survivability in the environment

Genetic engineering helps to design defective strains in a much more effective way than the classical methods.
On the basis of predictable effects, recombinant methodologies can be used at the best to limit long-term predictable effects. Predictability essentially is knowledge-dependent, i.e. the knowledge of the target and of the vector.

MONITORING

Monitoring techniques are available to test the tropism, the recombination potential and the survival in the environment. The detection and identification methods are based on nucleic acid sequence.

MAIN CONCLUSIONS

Similar vaccines can be designed either by the classical methods or using recombinant DNA technology. In that sense, there is no environmental problem specifically related to the use of recombinant DNA technology. However, genetic modification is sometimes the only way and very often the most efficient way to construct vaccines.

Main conclusions:

1. Knowledge can lead to familiarity, predictability, confidence and safety.
2. Deletion and insertional mutagenesis may lead to more knowledge.
3. Genetic engineering is often the best way to make vaccines as safe as possible from both short-term and predictable long-term effects.
4. Genetic engineering is making use of knowledge for the design and the production processes of vaccines.
5. Long-term effects are related to the goal(s), the target and the survivability of the vaccine but not to the technology. Should a vaccine be produced exclusively by genetic engineering, it follows that long-term effects are indirectly related to the technology.
6. Monitoring must be and can be done in relation to predictable effects.
7. Classical vaccines are not submitted to an environmental safety assessment, although the short-term and long-term ecological impacts of classical and recombinant vaccines are not perceived as different.

TOPIC 5

FISH

ECOLOGICAL EFFECTS OF GENETICALLY MODIFIED FISH: AN OVERVIEW

Dr. C. Pla (rapporteur), Laboratory of Ichthyology (Genetics), University of Girona, Spain

SUMMARY

Biotechnology has provided the appropriate tools for the building of new DNA molecules through the union of sequences from completely different sources. Genetically modified organisms or organisms produced entirely through the use of this technology are given the name "transgenics" and, the techniques commonly accepted as being likely to cause such phenomena are: (i) direct injection of the genetic material (into the pronucleus or into the cytoplasm of the fertilized ovum), (ii) cloning and transference of genes using vectors, and (iii) fusion of cells using non-naturally occurring techniques. The characteristics of high fecundity and external fecundation in the majority of fish species, has made this organism an attractive one for many research groups. Various genes have been used in these transgenesis experiments, covering a wide range of characteristics.

The utilization of transgenic fish may have an undesirable effect on the ecosystems if releasing of these organisms into the natural medium is produced. At this moment, there is no proof which can indicate to us with precision the possible effects which genetic modifications may have upon the fish itself. The only real danger perceived up to now, is the production of an organism with strange genetic material, but which is no way impeded from interbreeding with other populations of its species or with other different species. In the absence of data relative to transgenic fish, it is possible to use data obtained from non-modified fish populations as a model to study the potential effects of the transgenic organisms. In this sense, we present the restocking program used in several Spanish rivers by brown trout populations of foreign origin. The result of these repopulation schemes has been the progressive introgression of genetic material from the foreign populations with the consequent loss of the corresponding alleles of the native Spanish populations and, therefore, the loss of genetic diversity. Another consequence, has been the reduction in the size of the indigenous populations deriving from the competition with hatchery stocks.

INTRODUCTION

The rapid world-wide growth in the breeding of fish for consumption as food, has led to great effort in genetic research in order to improve the productive yield of various species which are of commercial interest. Throughout the past decade, the major part of this research work has concentrated upon the application of diverse gene manipulation techniques with the aim of improving the quantitative and qualitative yield of these species.

The induction of triploidy was the first technique to be used, in order to obtain sterile individuals, because of the interrelation which this has with the characteristics of their development. More recently, work has been carried out in the areas of gynogenesis and androgenesis, with the aim of modifying the sex ratios of the stocks.

However, the individuals obtained through these methods cannot really be considered to be genetically modified organisms (GMOs). Gene manipulation is always carried out between individuals from the same species and, in theory, these modifications could also occur naturally through random chance, from the crossing of different individuals from the same population; although, in practice, and over a limited period of time, the probability of this occurring spontaneously can be said to virtually zero. In the same way, the modifications obtained through *in vitro* fertilization, are not considered to be techniques which give rise to genetically modified organisms.

Modern biotechnological techniques have represented a great step forward in the field of gene manipulation since they permit transgression of the natural boundaries existing between different species. Biotechnology has provided the appropriate tools for the building of new DNA molecules through the union of sequences from completely different sources. The resulting product is often described as "recombinant DNA" and, the techniques employed are referred to colloquially, as "genetic engineering". The genetic material of a living organism can be isolated and transferred, either intact or in an altered stated, to various other kinds of organisms which differ greatly from one another. This transference may give rise to new recombinant organism, which possess determined extra genetic material of external origin. This can interfere with the development of certain hereditary characteristics, or can cause new ones to appear. Genetically modified organisms or organisms produced entirely through the use of this technology are given the name "transgenics" and, the techniques commonly accepted as being likely to cause such phenomena are: (i) direct injection of the genetic material (into the pronucleus or into the cytoplasm of the fertilized ovum), (ii) cloning and transference of genes using vectors, and (iii) fusion of cells using non-naturally occurring techniques.

PRODUCTION OF TRANSGENIC FISH

Since 1982, when the group of Palminter and Brinster presented their work on the mouse (Palminter *et al.*, 1982), demonstrating convincingly the application of these techniques, the production of transgenic animals has become a reality. The application of these techniques in order to produce transgenic domestic livestock mid-way through the '80's (Hammer *et al.*, 1985), has improved upon the results obtained through the classical reproduction techniques which were formerly used by breeders. In these processes, the manipulation is performed on selected individual genes rather than on the whole of an organism's genome, and has been limited to the addition of extra genes to the germinal line.

The characteristics of high fecundity and the presence of external fecundation in the majority of fish species, has made this organism an attractive one in the eyes of many research groups for the purpose of producing transgenic animals (Chorrout *et al.*, 1986 and Maclean *et al.*, 1987, on rainbow trout; Fletcher *et al.*, 1988 and Rokkones *et al.*, 1989 on salmon; Brem *et al.*, 1988 on tilapia; Zhu *et al.*, 1985, Yoon *et al.*, 1990 and Xie *et al.*, 1993 on goldfish and loach, amongst others). Various genes have been used in these transgenesis experiments, and different degrees of success have been obtained. The diversity of genes used covers a wide range of characteristics such as: growth hormone, induction of heavy metals, resistance to disease, antifreeze proteins and, regulating sequences, to name but some.

We shall not go into detailed description of the various techniques used, and/or likely to be used in the production of transgenic fish, their characteristics or the advantages and disadvantages of each of these with respect to the organism which we are describing. This description is going to be the object of a specific study in this session dedicated to fish, and will be carried out taking as a model the case of the salmon. We will only quote the essential steps to be followed at the moment of undertaking to produce a transgenic fish: (1) selection of the gene which interests us and construction of the corresponding genetic sequences flanked by the appropriate regulating sequences, (2) collection of fertilized ova, (3) microinjection of the genetic material into the fertilized cytoplasm, (4) recovery of the injected ova and testing of the transference, and, (5) analysis of the F1 and F2 generations.

APPLICATION AND FUTURE OF TRANSGENESIS IN FISH

Genetic engineering applied to the production of transgenic fish, is a subject of great interest concerning both the realization of basic research studies in the field of biology and also the genetic improvement of the species cultivated.

Control of reproduction, resistance to diseases and increase in production, whether due to reduction in the time of development or increase in the size of the individual, have been the principal objectives of the great majority of fish-breeders. However, in the great majority of breeding, the classical techniques of selection and reproduction have not achieved great improvements in the results obtained through these years. The impossibility of controlling the reproductive period in many marines species has led to the abandoning of specific breeding of these species, with the consequent stagnation in world production that this brings. This has diminished the expectations of several worldwide organizations, who believed that the production of marine cultures would have great repercussions with respect to the effort put into fishing all over the world. Knowledge and recognition of the genes responsible for these characteristics can, however, be applied in the obtaining of domestic transgenic stocks which favour the improvement of these cultures.

On the other hand, the production of transgenic organisms with specific genetic markers, could be used in the near future to solve ecological problems such as the migration of certain species of small pelagics. Transgenesis could also be applied to programmes of conservation of biodiversity, through the restoring, in certain populations, of naturally-occurring genes which, for one reason or another, have been lost.

EFFECTS ON THE ECOSYSTEMS

The utilization of transgenic fish may have an undesirable effect on the ecosystems if, for whatever reason, deliberate or not releasing of these organisms into the natural medium is produced. In the case of salmon, copious literature exists giving examples of escapes of these fish from the fish farms (Gausen & Moen, 1991). In order cases, the release of fish is carried out deliberately. This happens in many European countries as part of river repopulation programmes, using stocks of rainbow trout or brown trout from hatcheries.

The second case in this session will present a study of the possible ecological and genetic effects on the Atlantic salmon of the modifications introduced in its genome relative to the genes of the growth hormone and antifreeze proteins. Briefly, and as an introduction to this study, we will proceed to discuss the principal problem that the release of modified fish can cause in an ecosystem and the effects that may derive from this.

In the absence of data relative to transgenic fish, we must reaffirm our opinion with respect to the environmental effects of liberation of these, whether these effects be causal or not, working from our experience of populations of non-modified brown trout. The study carried out by our laboratory on the natural Spanish populations of brown trout, has shown a great genetic differentiation from those populations which are maintained in hatcheries. The latter are of foreign origin (mainly central European) and are used exclusively for the repopulation of rivers by different Spanish autonomous administrations. The result of these repopulation schemes, carried out on a massive scale, year after year, has been the progressive introgression of genetic material from the foreign populations with the consequent loss of the corresponding alleles of the native Spanish populations (García-Marín *et al.*, 1991). The first and gravest consequence of this introgression has been the loss, in some populations, of unique allelic variants, and, therefore, the loss of genetic diversity. The second consequence, deriving from the competition which these annual repopulations and the existence of diseases in the hatchery stocks represent, has been the reduction in the size of the indigenous populations. This reduction has reached dramatic proportions in those ecosystems with a high level of environmental pollution. At the present time, the indigenous Spanish populations of brown trout have become confined to stretches of the rivers which are close to the rivers' source.

At this moment, there is no proof which can indicate with precision the possible effects which genetic modifications may have upon the fish itself. Therefore, we must agree that the only real danger perceived up to now, is represented by the production of an organism with strange genetic material which is in no way impeded from interbreeding with other populations of its own species. And in the case of salmonids, with other different species.

In this way, we could recall at this juncture that the Ecological Society of America has suggested that, "the release of a genetically modified organism into a natural habitat be evaluated bearing in mind its biological properties (phenotype), rather than the genetical techniques used to produced it" (Tiedje et *al.*, 1989).

Thus, in the case of the production of domestic stocks for use in fish culture, the introduction of a gene (or genes) which provoke the sterility of the organism should be demanded, in order to minimize as far as possible the impact which accidental releases into the natural habitat may represent. If this is not done, the effects which have been produced so far with the domestic stocks of different origin from the local populations in the areas where the hatcheries are situated, could be reproduced on a more intense and wider-reaching scale. Therefore, the next step to be taken would be to stimulate genetic research in the direction of identifying and transferring genes which are related to the fertility of the organisms, before beginning massive production of domestic stocks through transgenetic processes. This should also be applicable to any kind of domestic stocks possessing sufficiently different genetic information from that of the local populations of its species. If this is not done, we will be opposing the internationally supported criterion of advancing in the field of the preservation of present biological diversity.

With respect to the cases of production of transgenics for deliberate release, such as those possible experiments to be carried out in specific ecosystems mentioned earlier, these should be strictly controlled until it was absolutely certain that the modification introduced does not produce, in itself, any harmful effects on the individual which has undergone this modification.

REFERENCES

Brem, G., Brening, B., Horstgen-Schwarrk, G. and Winnacker, E.L. (1988): Aquaculture, 68: 209-219.

Chourrout, D., Guyomard, R. and Houdebine, L. (1986): Aquaculture, 51: 143-150.

Fletcher, G.L., Shears, M.A., King, M.J., Davies, P.L. and Hew, C.L. (1988): Can. J. Fish. Aquat. Sci, 45: 352-357.

García-Marín, J.L., Jorde, P.E., Ryman, N., Utter, F. and Pla, C. (1991): Aquaculture, 95: 235-249.

Gausen, D. and Moen, V. (1991): Can. J. Fish. Aquat. Sci., 48: 426-428.

Hammer, R.E., Pursel, V.G., Rexroad, C.E., Wall, R.J., Bolt, D.J. Ebert, K.M., Palminter, R.D. and Brinster, R.L. (1985): Nature, 315: 680-683.

Maclean, N., Penman, D. and Zhu, Z. (1987): BioTechnology, 5: 257-261.

Palminter, R.D., Brinster, R.L., Hammer, R.E., Trumbauer, M.E., Rosenfeld, M.G.,

Birnberg, N.C. and Evans, R.M. (1982): Nature, 300: 611-615.

Rokkones, E., Alestrom, P., Skjervold, H. and Gautvik, K.M. (1989): J. Comp. Physiol. B, 158: 751-758.

Tiedje, J.M., Colwell, R.K., Grossman, Y.L., Hodson, R.E., Lenski, R.E., Mack, R.N. and Regal, P.J. (1989): Ecology: 70: 298-315.

Xie, Y., Liu, D., Zou, J. Li, G. and Zhu, Z. (1993): Aquaculture, 111: 207-213.

Yoon, S.J., Hallerman, E.M., Gross, M.L., Liu, Z., Schneider, J.F., Faras, A., Hackett, P.B., Kapuscinski, A.R. and Guise, K.S. (1990): Aquaculture, 85: 21-33.

Zhu, Z., Li, G., He, L. and Chen, S. (1985): Z. Angew. Ichthyol. , 1: 32-34.

GENETICALLY MODIFIED FISH: TECHNOLOGIES AND THEIR POTENTIAL ENVIRONMENTAL IMPACTS

Dr. D. Chourrout, Laboratoire de Génétique des Poissons, Institut National de Recherche Agronomique, France

SUMMARY

Development of modern fish farming has led to overproduction which oblige to reduce the production costs. Part of the solutions lies in genetic improvement of domestic stocks, which enhances the genetic divergence between farmed fish and wild fish of the same species. Physical containment of captive fish has proved impossible in most cases, and introgression into wild stocks is therefore a major concern, although mass restocking operations (deliberate release of domestic fish) does not always lead to such phenomena. Genetic modifications can be subdivided in chromosome set manipulations (induced uniparental reproduction and polyploidy) and transgenic technologies. Uniparental reproduction leads to accelerated inbreeding in a way similar to conventional methods of breeding, and to unisexual populations identical to those resulting from hand-sorting of both sexes. Triploidy leads to desirable sterile fish, which offers an effective tool of biological containment to avoid introgression into wild stocks. Tetraploidy may present a danger in case of accidental release, but tetraploid fish have not yet to be produced on a large scale. Stable lines of transgenic fish can be produced by several methods, leading to integration and expression of foreign genes. Traits to improve by gene transfer are often those chosen for selective breeding, and escaped fish obtained with either methods may have similar effects on natural populations, provided that gene transfer does not modify other functions at the same time. Another major uncertainty is the extent to which traits improved for captive life will change the fitness in nature. Improved cold tolerance by antifreeze gene transfer may be an exception in this respect, and should be considered with a particular care.

INTRODUCTION

Fish farming, which is a traditional activity in various regions of the world, has been strongly developed in the last two decades, due to the plateau attained by fisheries catches. In rich countries, this development is stimulated by an increased demand for high value species, which are particularly overexploited by fisheries. The major achievements of modern fish farming concern salmonids grown in freshwater and then in seawater cages (Norway, Scotland, Chile), a number of strict marine species in the Mediteranean sea and Far East, and the freshwater channel catfish in North America. After their development, these activities have reached levels of overproduction, and it is now necessary to reduce production costs, either by improving the farming conditions or by using genetically improved fish stocks.

Only few countries up to now have been able to undertake large-scale programs of fish genetic improvement. These programs and many results of fish genetics research show that classical selective breeding is often adequate for traits having a major importance for fish farming productivity (growth rate and disease resistance). More recently, it was also shown that genetic modifications such as uniparental reproduction and polyploidy could be part of fish improvement programs, although they had been used essentially for crop production in the past. Genetic modifications involving gene manipulations have also been addressed in fish, in purpose of gene regulation studies but also as a promising tool of genetic improvement. It is clear, as it is for other organisms, that introduction of genetic modifications into fish raises many concerns for the consumption but also for the environment. In this article, we will describe the state-of-the-art in fish genetic manipulations and their potential for genetic improvement. Each of them deserves specific comments about their ecological impacts in case modified fish are released into the environment. However, we would like to introduce first the conditions in which this release already takes place with non-modified fish, and the major questions generated by contamination of wild stocks by domestic fish.

INTERACTIONS BETWEEN DOMESTIC AND WILD POPULATIONS

Fish farms communicate physically with their environment through natural freshwater or seawater. Wild populations of the same species or of closely-related species may live in the surroundings. Both worlds are separated by mesh or grid barriers, and domestic fish constantly escapes from their farms. The chance to recapture a significant proportion of them is negligible. They diverge from their wild counterpart by their own history but also genetically, because they do not generally result from wild stocks of the same area, which have been submitted in captivity to distinct pressures of stock management including unconscious or eventually conscious selection. Introgression of wild populations can happen in a rather unpredictable manner, as this is currently shown by using pertinent genetic markers (Barbat-Leterrier et al., 1989). There is a debate on the extent to which introgression works, which is stimulated by quite variable observations, depending upon the species, the stocks in presence, and the geographical areas. However, introgression is sufficiently documented to trigger another major debate on its qualitative impact on natural gene pools. This discussion opposes two extreme attitudes: one opinion is that domestic fish release represents a helpful quantitative input for the species, and that introgression reveal a good adaptation of intruders to the local conditions. This justifies mass restocking operations (deliberate release) with domestic fish; the other opinion is that any input from farmed stocks weakens wild populations, because farmed fish will compete, dominate and genetically pollute them, with long-term negative effects on wild stocks (this suggests that intruders are always less adapted to nature than their natural counterparts). Unfortunately, these two attitudes essentially differ by their degree of pessimism. There is no rational argument available yet, to qualitatively measure the impacts of domestic fish release. Therefore, it is worth preventing as much as possible the introgression by physical or biological containment, especially when domestic and wild stocks strongly diverge. Genetic modifications increase this divergence, in a way that we will attempt to discuss case by case in the following sections.

GENETIC MODIFICATIONS OF FISH

We will not treat here the modifications achieved by conventional selective breeding, although they represent today the major source of divergence between domesticated and wild stocks of the same species. They obviously confer better performances to fish in captivity (Gjedrem, 1983), which may also be expressed in the natural environment in case of release. Consequently, there is no reason to minimize the impact (eventually negative) that selective breeding may have on the environment. We will address here two major types of genetic modifications. Uniparental reproduction and polyploidy involve the same technologies, which modify the chromosome complement (chromosome set manipulations). Gene manipulations are up to now restricted to transfers of individual foreign genes into fish genome.

Chromosome set manipulations

Most fish species have separated sexes, and diploid embryos result from the fusion of haploid sperm and diploid oocytes. Half of the diploid oocyte chromosome set is expelled during the completion of meiosis II, and the resulting haploid female pronucleus fuses with its haploid male counterpart to form the nucleus of the first cell, ready to enter the first mitotic cycle. Uniparental reproduction consists of an elimination of either male or female chromosome set (gynogenesis and androgenesis). Polyploidy consists of an addition of chromosome sets to the diploid complement (Ibsen et al., 1990 ; Thorgaard et al., 1992).

Gynogenesis and androgenesis

To induce gynogenesis, the paternal genome is usually eliminated by sperm irradiation prior to in vitro fertilization. Ultraviolet irradiation is preferred to treatment with ionizing rays, for safety reasons and also because it leads to a total elimination of male chromosomes for a larger range of doses. After fertilization, haploid embryos start to develop and usually die before or just after egg hatching. Production of viable gynogenetic individuals requires the diploidization of maternal chromosome set, which is achieved by an inhibition of chromosome disjunction during either meiosis II or the first embryo mitosis. Thermal or pressure egg treatments are very efficient to suppress these two disjunctions. For androgenesis, the oocyte is generally irradiated with ionizing rays in order to destroy the female genome, and the male chromosome set is doubled by preventing the first mitotic disjunction. Androgenetic and gynogenetic individuals which result from mitotic inhibition have in general very low survival and growth rates, certainly because they are homozygous at all their loci. Gynogenetic fish issued from meiotic interference are also poorly performant, but more viable, presumably because they remain heterozygous at many loci. This is one reason for which these genetically modified fish are never directly used for farming.

Gynogenesis and androgenesis allow a rapid production of inbred strains, which are then crossbred to obtain a hybrid vigor. Gynogenesis is also used, because it eliminates the Y chromosome in "female homogametic species" (XX-XY) and provides genetic female offsprings exclusively. Females are more interesting than males in many types of fish farming, because their puberty occurs later and at a larger size; sexual maturation is not desirable in fish because it involves a tremendous energetic cost devoted to the production of ripe gonads. Small numbers of gynogenetic juveniles are sex-reversed by hormonal treatments during the course of sex differentiation, and resulting males are then mated with large numbers of standard females to provide all-female outbred offsprings.

It is noteworthy that gynogenetic, androgenetic fish, and their outbred offsprings do not differ in their genetic nature from normal individuals, although they are either specially inbred or characterized by abnormal sex ratios. Their features could be obtained by long and tedious sib-mating or by conventional methods of sex control (including hand-sorting of both sexes as soon as they are discernible). For these reasons, we will not consider here that these two methods can have a specific impact on the environment, in case their direct or indirect products contaminate natural populations.

Polyploidy

There are at least two viable forms of polyploidy in fish, induced in eggs where both paternal and maternal are maintained. Triploidy is induced very frequently on a large scale, by mass treatment of fertilized eggs with temperature, shortly after a normal fertilization. As mentioned above, this operation doubles the female chromosome set, which is then associated with a haploid number of male chromosomes. A later treatment results in tetraploidy, via the inhibition of first mitotic disjunction. Triploids are in general highly viable and slightly less performant than diploids until sexual maturation.

Triploid gametogenesis is then deeply reduced, especially in salmonid females which are totally sterile. This makes their greatest interest for the production of large-size fish, since sterility is correlated with increased growth rate and meat quality. Male triploids are eliminated by a simple combination of triploidization and female monosexing. This method is currently used in long term salmonid farming, except in Norway where selection for late sexual maturation has been chosen and kept as the only method of sex control. However, we can predict that triploidy will be the method of choice in all possible cases, since triploid fish escaping from farms cannot introgress natural populations. Therefore, triploidy is a capital tool to minimize genetic impacts of farmed stocks on the environment.

Tetraploidy is up to now of more interest, and has been generated with two main purposes. Because tetraploids can accomodate up to four different alleles at each loci, improvements of major traits could be attained by their production and subsequent reproductions. A large scale production of tetraploids would have no sense, except if they are rendered more performant than diploids. However, tetraploids of first generation have shown very low performances in salmonids, and only a long breeding investment may result in such progress. At least theoretically, we may expect a real danger in case large numbers of tetraploids would be in a position of escapement from conventional farms; as a matter of fact, they would eventually mobilize the gametes of wild diploid partners to generate infertile progenies. The tetraploid ability to generate triploids when they mate with diploids has been demonstrated in salmonids. It could have been used on a large scale to induce sterility. However, direct production of triploids from diploid parents has been preferred up to now, for a number of technical reasons. We have to mention that few tetraploid adults have to be grown up to generate large numbers of triploids.

Technologies of transgenic fish production

After the pioneer works in transgenic mice, several groups have initiated experiments to transfer foreign genes into fish embryos. The high fecundity, external fertilization and development, were good advantages for any egg manipulations including injections of DNA at early stages. On the other hand, difficulties were the availability of few cloned DNA sequences isolated from fish genome, and that fish were not amenable to the technologies designed for transgenic mice production. Pronuclei are not visible in the eggs of any species, and no retrovirus possibly at the origin of efficient integration vectors have been identified in fish. The first attempts were focused on DNA injections into the egg cytoplasm, and other technologies have been tested more recently (Houdebine and Chourrout, 1991). After a description of these operations and their general consequences in terms of gene integration and expression, we will describe the gene transfer projects in progress to improve commercial traits of farmed fish, and discuss their eventual consequences in case of transgenic fish release.

Cytoplasmic injections

In first attempts, circular or linear DNA were injected into the cytoplasm of fertilized eggs, generally in the first cell, but sometimes after few mitotic cycles. Very large amounts of DNA are introduced, certainly in order to compensate the relative imprecision of cytoplasmic injections (million to billion copies). An intense replication of foreign DNA is observed, with a quick passage to high molecular-weight forms, and is followed by a massive degradation. This shows that most injected material has persisted transiently in extrachromosomal concatemers. For this reason, early analyses are not indicative of integration processes. Later examinations however permit to detect small amounts of foreign DNA, generally distributed mosaically in a minority of fish. Transmission is also observed to a minority of F1 offsprings, showing that the germline is also mosaic, and to mendelian proportions of F2 individuals. This clearly suggests that foreign DNA has been integrated, and convincing evidences for stable integration have recently been obtained with pulse-field gel electrophoresis and in situ chromosome hybridizations. In most cases, DNA is integrated in one or two sites of the genome, in single or multiple copies. Multiple copies may be assembled in very large concatemers, where they are oriented randomly. The extent of transgene rearrangement has not been thoroughly studied yet. To summarize, cytoplasmic injections permit the production of stable transgenic lines, in one more generation than does pronuclear injection in mouse.

Other routes of integration

A major other way for stable integration is the injection of smaller amounts of DNA into the nucleus of preovulatory oocytes (germinal vesicle), which are then matured and inseminated in vitro. This also leads to mosaic distribution. Germinal vesicle injection has been set up in two species (medaka and goldfish) in which it appears more efficient than cytoplasmic injection. Other attempts tend to reduce the mosaicism and to produce larger numbers of transgenics in a very short time. Sperm incubation with linear DNA has been unsuccessful so far, but sperm electroporation looks more promising. Egg electroporation appears to be the major alternative however, although its efficiency is not yet precisely known. In case it will prove satisfactory, the development of many new projects of fish gene transfer can be expected. Finally, interesting efforts towards potent integration vectors must be mentioned. Addition of sites recognized by retrovirus integrases in the injected construct seem to favourize earlier integration, when the specific integrase is cointroduced. Recent in vitro studies have also shown that mammalian retroviral vectors enveloped in the G protein of the wide host-range VSV virus can efficiently infect and transform fish cells.

Gene expression in transgenic fish

Earlier studies used gene regulatory elements originated from other organisms, since very few fish sequences had been cloned. Most of them may not be informative about stable expression, because gene expression was monitored at early development stages after injection. A number of later examinations concerned fish harbouring mammalian growth hormone genes coupled with ubiquitous promoters, in purpose of growth stimulation. The results are controversial, because authors who reported fish size increments had often not really proven that transgenes were expressed. A second phase of studies were performed with reporter genes (CAT in particular) in order to compare various promoters after cell transfection or embryo transformation. The good potential of several heterologous promoters was demonstrated, as well as the conservation of their original properties in fish cells (tissue-specificity, inducibility). Other heterologous genes, especially those containing their own introns and terminators, were not efficiently expressed with the same promoters. To understand these difficulties, many constructs were made, with heterologous promoters and reporter genes, in which the effects of various introns, leader and polyA sequences were compared in transient and stable transfection of several fish cell lines. Remarkable expression levels were observed for certain combinations, but optimization of a vector for general use seems difficult or impossible. Indeed, strong interactions were found between the effects of each vector elements, and a particular vector may be appropriate for one gene and inadequate for another.

Physiological modifications of transgenic fish and eventual consequences

Growth

Growth rate is a critical trait in modern fish farming, and has been addressed by gene transfer technologies. Impressive growth stimulations have been registered after transferring salmonid GH cDNAs coupled with RSV promoter (in carp and catfish; Zhang et al., 1990) or with fish antifreeze gene promoter (in salmon; Du et al., 1992). Concentrations of exogenous growth hormone in the plasma were very low if detectable, but the biological effect may have resulted from permanent secretion of little amounts of growth hormone. Transgenic fish with GH genes or genes coding for related mediators have not entered fish farming yet, and deserve to be tested in detail in comparison with standard improved strains. In terms of environmental impact, experiments are needed to know how larger fish will interact with wild populations. In principle, other fast-growing fish resulting from selection might have the same impacts, but we must keep in mind that GH affects other functions than growth (smoltification in particular). The abnormal regime of exogenous GH secretion in transgenics may have peculiar effects on these functions, which will not occur with selected fish.

Cold tolerance

In another major gene transfer project in salmonids is cold tolerance envisaged by the expression of antifreeze protein genes isolated from arctic and antarctic fish species (Fletcher et al., 1988). A number of areas on the Northeastern Canadian coast are appropriate for salmon farming, except during the winter, because seawater freezing obliges to transfer the fish back into freshwater. Antifreeze protein genes are organized in clusters in their original configuration, and a part of them has only been transferred into salmon embryos. They were expressed at a very low level, far under the minimum required for a physiological effect. Solutions for the future may reside in the transfer of much longer genomic clones, which is now rendered possible in transgenic mice. An important point is that such a cold tolerance is hard to expect from selective breeding. Another noteworthy aspect of this project is the risk of major ecological effects in case of trangenic fish release: an extension of the salmon distribution area could result from their increased cold tolerance.

Disease resistance

Other projects also trigger particular concerns, because the features of transgenic fish may give them a higher fitness in the environment. Transfer of viral genes or specific immunoglobulins is envisaged to improve disease resistance, which is also a critical aspect of farming profitability. Selective breeding seems to have a reasonable potential in this direction. We must underline that specific pathogens are particularly influent on fish survival in intensive rearing conditions, but may play much less role in wildlife. An advantage of gene transfer strategies may be that the resistance can remain strictly pathogen-specific, in contrast to conventional selective breeding which may enhance a wider resistance by stimulating non-specific defences.

CONCLUSIONS

To summarize, genome modifications have been set up in fish with the purpose of genetic improvement for farming. Transgenic fish are under major scrutiny for the question of ecological impacts, although technical limitations delay their entrance in fish farms. Traits to improve by gene transfer are often those chosen for selective breeding, and escaped fish obtained with either methods may have similar effects on natural populations. Another major uncertainty is the extent to which traits improved for captive life will change the fitness in nature. Improved cold tolerance by antifreeze gene transfer may be an exception to be considered with a particular care.

If genetic modifications pose specific problems in terms of environment, genetic pollution is basically possible because farmed stocks genetically diverge from natural populations and because total physical containement is non-existant in fish farming. Introgression into natural populations is not certain a priori, even when domestic fish are deliberately released for restocking purposes. Introgression happens however and is certainly correlated with the biomass level of domestic intruders. The only present and rather satisfactory solution to avoid introgressions is the biological containment offered by induced triploidization. It is very interesting to note that triploidy has been set up for fish farming productivity, and may secondarily offer solutions for environmental protection.

REFERENCES

Barbat-Leterrier A., Guyomard R., Krieg F. (1989) Introgression between introduced domesticated strains and Mediterranean native populations of brown trout (Salmo trutta L.). Aquat Liv Res 2 : 215-223.

Du S.J., Gong Z., Fletcher G.L., Shears M.A., King M.J., Idler D.R., Hew C.L. (1992) Growth enhancement in transgenic Atlantic salmon by the use of an "all fish" chimeric growth hormone gene construct. Biotechnology 10 : 176-181.

Fletcher G.L., Shears M.A., King M.J., Davies P.L., Hew C.L. (1988) Evidence for antifreeze protein gene transfer in Atlantic salmon (Salmo salar). Can. J. Fish Aquat. Sci. 45 : 352-357.

Gjedrem T. (1983) Genetic variation in quantitative traits and selective breeding in fish and shellfish. Aquaculture 33 : 51-72.

Houdebine L.M., Chourrout D. (1991) Transgenesis in fish. Experientia 47 : 891-897.

Ihssen P.E., McKay L.R., McMillan I., Phillips R.B. (1990) Ploidy manipulation and gynogenesis in fishes. Trans. Am. Fish. Soc. 119 : 698-717.

Thorgaard G.H., Scheerer P.D., Zhang J. (1992) Integration of chromosome set manipulation and transgenic technologies for fishes. Mol. Mar. Biol. Biotech. 1 : 251-256.

Zhang P., Hayat M., Joyce C., Gonzalez-Villasenor L.I., Lin C.M., Dunham R.A., Chen T.T., Powers D.A. (1990) Gene transfer, expression and inheritance of pRSV-rainbow trout-GH cDNA in the common carp, Cyprinus carpio (Linnaeus). Mol. Reprod. Dev. 25 : 3-13.

ECOLOGICAL AND GENETIC EFFECTS OF TRANSGENIC FISH

Dr. K. Hindar, Norwegian Institute for Nature Research, Norway

SUMMARY

This study considers the long-term ecological and genetic effects of using genetically modified fish in aquaculture. In doing so, I rely on experience from introductions of non-native species and populations of fish that have occurred for more than a century. Throughout, the emphasis is on salmonids and in particular on Atlantic salmon (*Salmo salar*). Releases of non-native (including cultured) salmonids into natural settings have resulted in a wide variety of outcomes, ranging from no detectable effect to complete introgression or displacement. Where genetic effects on performance traits have been documented, they appear always to be negative in comparison with the unaffected population. These observations raise concerns over the genetic future of many natural populations in the light of increasing numbers of released fish, even in the absence of genetic modification. The genetic modifications which appear likely to become realized in salmon aquaculture are the introduction of growth hormone genes and antifreeze protein genes into salmonid genomes. I discuss three scenarios for the culture of Atlantic salmon carrying these genetic modifications. The first scenario assumes that transgenic fish will be used in aquaculture according to their profitability, and that no measures are taken to reduce the number of fish escaping from aquaculture. In this scenario, current negative trends for natural populations will be reinforced. In particular, the use of transgenic cold-tolerant Atlantic salmon would lead to a northwards expansion of aquaculture, so that a number of fish populations that are now spared from the effects of escaped fish would be at risk. In a broader taxonomic perspective, northwards expansion of aquaculture could lead to irreplaceable loss of biological diversity caused by introduction of a novel predatory fish to northern aquatic ecosystems. The second scenario assumes that concerns about the effects of genetically modified fish lead to better physical containment than in present-day fish farms. This would reduce interactions between escaped and wild fish, but would still have the potential for dramatic effects in cases where technical failure or human error led to large-scale escapes. The third scenario assumes that gene technology be used actively in search for biological containment of transgenic fish, in addition to physical containment. Only in this third scenario is the use of transgenic fish in aquaculture compatible with internationally established goals for conservation of biological diversity.

INTRODUCTION

There is considerable interest worldwide in applying gene technology to aquaculture. It is hoped that an increasing demand for marine proteins can be provided by aquaculture, and that genetic modification can be used to tailor aquaculture species for cost-efficient production of proteins. The application of gene technology in aquaculture lags several years behind applications in agriculture. Nevertheless, recent developments in gene transfer to fish indicate that transgenic fish may have a future in commercial aquaculture, just like transgenic cultural plants are now being increasingly used in agriculture. Unlike most cultured plants, transgenic fish are not intended for release but for captive rearing to market size. Experience from current aquaculture, however, shows that accidental releases are common. This mandates a thorough evaluation of the environmental effects, should the transgenic fish escape.

The environmental effects may be divided into three broad categories (Williamson et al. 1990): (1) effects caused by the genetically modified organism itself, (2) effects resulting from dispersal of genes from the genetically modified organism to other organisms in the environment, and (3) altered practice in the use of an organism because of the genetic modification. Experiments with transgenic fish are too few to act as a knowledge base for assessments of their potential environmental effects (and to my knowledge, none of those experiments were designed to test hypotheses about environmental effects). This means that we have to rely on experience from releases of non-modified fish in order to predict the consequences of genetically modified ones. This study is based on considerations of the environmental effects of releases of (non-modified) salmonid fishes, which can serve as excellent models in this context. Salmonids are a target for genetic modification, and information from this group provides one of the best data sets for evaluating the ecological and genetic effects of releases of non-native populations to aquatic environments (Hindar et al. 1991).

The presentation is in five parts. First, I make predictions about how transgenic fish may be used in aquaculture of Atlantic salmon (*Salmo salar*) in the future. Second, I review experience gained about the ecological and genetic effects following releases of non-native salmonids. Third, I combine knowledge about gene technology and effects of releases to speculate about the effects of transgenic fish. Fourth, I discuss three scenarios for the future use of transgenic fish in salmon aquaculture, and evaluate whether or not they are compatible with internationally established goals for conservation of biological diversity. Finally, I make conclusions about the use of transgenic fish in aquaculture. The reader is referred to Hindar (1993) for a more detailed account of the presentation.

GENETIC MODIFICATION OF FISH

A detailed account of the techniques and prospects for genetic modification of fish is presented by Chourrout (this volume). In salmon aquaculture, it appears that two kinds of genes are particularly interesting. First, there is an interest in producing transgenic strains of fish that express higher levels of growth hormone (GH), and, once they have been generated, transmit the enhanced growth rate to the next generations. Transgenic fish microinjected with growth hormone genes have now been shown to grow faster than the controls. Moreover, transgenic progeny of these fish grow faster than their non-transgenic siblings, and also much faster than their parents.

Second, genes that provide tolerance to cold water have economic interest in aquaculture, because the success of the marine phase of salmon aquaculture is in many areas restricted by cold winter temperatures (e.g. the Atlantic coast of USA and Canada, and the Barents Sea coast of Norway and Russia). Especially problematic is the winter survival in net pens in the supercooled sea water which occurs in these areas. During winter time, seawater temperatures in the Arctic may go down to -1.4 to -1.9 °C. Atlantic salmon cannot tolerate the cold water and freeze to death at temperatures lower than -0.7 °C. However, other fish species are adapted to extremely low water temperatures by the evolution of antifreeze genes which encode proteins that keep their blood from freezing. A number of experiments have been carried out to transfer antifreeze protein (AFP) genes to Atlantic salmon. These experiments showed integration of the antifreeze genes into the salmon genome and evidence of antifreeze protein expression, although at levels that are 100-fold less than would be required to provide freezing protection. It is probably only a question of time until adequate antifreeze protein expression is achieved in order for this transgenic system to be of commercial interest.

In theory, captive rearing does not result in releases of cultured organisms into the wild. However, the number of accidentally released fish from aquaculture operations is considerable; in Norway, the biomass of escaped fish surpasses the annual yield of wild fish in the sea and rivers combined. As a consequence of the escapes, more than 30% of the Atlantic salmon spawners in Norwegian rivers are of farmed origin. This requires that the use of transgenic fish in aquaculture must be evaluated as potential release experiments.

ECOLOGICAL AND GENETIC EFFECTS OF NON-MODIFIED FISH

Here I summarize the experience gained from one hundred years of deliberate and accidental release of non-modified salmonids, based on reviews focusing on effects of releases at interspecific (mainly 'ecological') and intraspecific (mainly 'genetic') levels (Hindar et al. 1991, Krueger and May 1991). Disease introductions and effects on population size are reviewed in separate sections, because they may have both ecological and genetic causes.

Ecological effects

A number of introductions of salmonid species outside their native range have taken place both between and within continents. Self-sustaining populations of North American salmonids, most notably rainbow trout (*Oncorhynchus mykiss*) and brook trout (*Salvelinus fontinalis*), have been established in Europe, South America, Africa, Asia and Australia. Likewise, the European brown trout (*Salmo trutta*) has been introduced to all continents except Antarctica, and Atlantic salmon is now introduced to new continents for aquaculture. Introductions of brown trout appears to be the major factor causing declines and disappearance of many brook trout populations in eastern North America. Apparently, brown trout excludes brook trout from preferred feeding and resting positions in streams by interference competition (Krueger and May 1991).

Knowledge about the mechanisms by which one species replaces another may be useful when extrapolating these observations to speculations about transgenic fish. In streams, competition occurs as species attempt to secure territories for adequate space, food and cover. Such interference competition is primarily mediated through aggressive behavior towards other individuals. Brown trout appear to be the most aggressive of the salmonids, more than the congeneric Atlantic salmon, and more than all of the Pacific salmon (*Oncorhynchus*) and charrs (*Salvelinus*). Size-related competitive ability may be another mechanism by which one salmonid species gains a competitive advantage over another, probably through a more efficient exploitation of food resources.

Genetic effects

Two broad conclusions regarding genetic effects on native salmonid populations emerge from a review of the empirical observations following releases of non-native salmonids to the environment (Hindar et al. 1991):

1. The genetic effects of (intentionally or accidentally) released salmonids on natural populations are typically unpredictable; they vary from no detectable effect to complete introgression or displacement.

2. Where genetic effects on performance traits have been detected, they appear always to be negative in comparison with the unaffected native populations.

The typically negative effects of releases of non-native salmonids are not unexpected. Salmonid populations are adapted to their local environments (Taylor 1991), and any introduced populations or crosses involving introduced populations should be expected to perform worse than the native ones.

Recent evidence suggests that the genetic effects may go beyond the intraspecific level. Elevated rates of hybridization between Atlantic salmon and brown trout have been observed in Norwegian rivers having high proportions of escaped farmed Atlantic salmon (Hindar & Balstad, Norwegian Institute for Nature Research, unpublished results). It is not yet known whether this will lead to introgression between the two species.

Disease introduction

Diseases are a serious problem in salmonid culture throughout the world, and represent a threat to many salmonid populations in nature. Transfer of fish between culture operations has led to pathogens and parasites coming in contact with fish populations to which they were not adapted. These new host-pathogen confrontations have led to unnaturally high fish mortality. In Norway, Atlantic salmon populations in 35 Norwegian rivers have been greatly reduced since 1975 following introductions of fish from hatcheries infected by the monogenean parasite *Gyrodactylus salaris*. This parasite does not appear to be native to Norway, and was probably imported with salmonid eggs or juveniles from infected localities in the Baltic. Another problem has been furunculosis caused by the bacterium *Aeromonas salmonicida*. This pathogen was introduced to Norwegian fish farms with infected smolts from Scotland in 1985, and had spread to more than 500 fish farms (65% of the total) and 63 rivers (15%) by the end of 1991. This latter example shows the enormous potential that aquaculture activities have for the spread of epidemics.

Effects on population size

A decline of local wild populations has been observed to accompany releases, in particular following disease introduction, introgression and/or displacement. In other cases the reasons for the decline remain obscure, but it is possible that both competition between released and wild fish, and increased predation rates (including overharvesting) contribute to the decline.

POTENTIAL IMPACTS OF RELEASING TRANSGENIC FISH

In this section, two types of transgenic salmonids - both of them already a reality in the laboratory - will be discussed with respect to the potential effects on wild salmonids. One is transgenic fish with inserted growth hormone genes for enhanced growth rate. The other is transgenic fish with inserted antifreeze genes for enhanced tolerance to cold sea water.

Ecological novelties

All of the effects that were summarized in the previous sections should be considered when predicting ecological effects of transgenic fishes. Moreover, transgenic fishes may pose greater ecological risks than escapes from conventional aquaculture because "organisms with novel combinations of traits are more likely to play novel ecological roles, on average, than are organisms produced by recombining genetic information existing within a single evolutionary lineage" (Tiedje et al. 1989).

According to Tiedje et al.'s description of 'ecological novelties', it is doubtful whether fish with inserted growth hormone genes qualify for that category or not. On the one hand, transgenic fish with extra copies of species-specific growth hormone genes would not be considered dramatically changed by gene technology. On the other hand, if their growth rates were beyond the range expressed by non-modified individuals of the same species, they could definitely play novel ecological roles upon release or escape. Since large body size is one mechanism by which one fish gains a competitive advantage over another, faster growth rates in transgenic Atlantic salmon would mean that they could be at an advantage both in intraspecific and interspecific contests in rivers. In the former case, this would affect native Atlantic salmon populations negatively; in the latter, it would affect brown trout and Arctic charr coexisting with Atlantic salmon.

Based on studies of growth hormone administration to fish, transgenic fish could also have higher appetite. The food requirements of the transgenic fish would therefore easily lead to starvation in periods of food shortage (Powers et al. 1992). If so, transgenic fish would experience poor survival in the many rivers where food availability limits growth rate, and most probably even in more productive rivers during periods of food shortage. The combined effect of a transgenic fish that on the one hand has periods of faster growth rate than native fish, and on the other hand has periods of higher mortality, would very much depend on when and how they entered the stream ecosystem. If they entered through the reproduction of transgenic fish, they would survive the start-feeding period if the timing of emergence matched the peak abundance of food organisms. Those surviving the start-feeding period would most probably find difficulties surviving the first autumn in unproductive rivers and would survive longer in productive ones. If, on the other hand, the transgenic fish entered the river through escapes from freshwater hatcheries, there would probably be larger effects on the river ecosystem, because there would always be new fish entering the river, and some of them would be very large-sized compared with wild fish.

Little or no information exists for transgenic fish on changes in other traits that are genetically or phenotypically correlated with growth rate (Kapuscinski and Hallerman 1991). At present, we have to rely on information gained from exogenous administration of growth hormone to fish. Growth hormone plays a role in osmoregulation in anadromous salmonids. During migration from fresh water to salt water, levels of growth hormone are elevated, leading to an increase in salt (Na+) exclusion at the gills. Transgenic, emigrating smolts would therefore be likely to avoid predation better than wild smolts upon entering salt water, because they would adjust faster to the saline environment and thereby escape estuarine and coastal predation. Two factors would contribute to the faster osmoregulatory adjustment of transgenic smolt to salt water; a higher level of growth hormone, and a larger body size.

Most fish that escape from fish culture today do so from net pens in the marine environment. Considering the extremely rapid growth of Atlantic salmon in salt water, I expect there would be only short-term problems with food shortage for transgenic fish, unlikely to lead to increased mortality. Thus, escaped transgenic fish would be at a selective advantage compared with wild salmonids in salt water. This suggests a generally negative effect on wild salmonids through intraspecific and interspecific competition. The effects on other marine species can only be guessed upon as long as we know so little about their interaction with Atlantic salmon.

Transgenic salmonids expressing antifreeze protein genes from winter flounder or any other species which possess such genes would definitely fall in the category of ‘ecological novelties’. The effects of transgenic cold-tolerant fish would most probably be largest on wild populations of Atlantic salmon and Arctic charr, and on marine fishes which express antifreeze proteins naturally. Antifreeze proteins would confer a selective advantage to salmon in any location where cold sea water constrains their life history today. In rivers draining to Ungava Bay, Canada, there are indications that the life cycle is curtailed through maturation in fresh water and no anadromous migration (Power 1969). A smolt expressing antifreeze proteins would be able to emigrate at a smaller size and a younger age, and gain weight (and fecundity) in the sea to spawn at a younger age than individuals of the local populations. Thus, one can easily envisage how transgenic Atlantic salmon could outcompete native populations in these rivers. Transgenic freeze-resistant fish could also contribute to a northwards expansion of Atlantic salmon. This would put Arctic charr populations at risk, especially river populations which are considered to be less competitive in that environment than most other salmonids, and which therefore thrive only in physically extreme environments as in the northernmost rivers. These effects would be further strengthened if the use of transgenic freeze-resistant fish moved aquacultural production to the north (see below).

Interbreeding

Genetically modified fish would be expected to result in the same genetic effects as non-modified ones, both with respect to changes in genetic structure and with respect to loss of genetic adaptations to local environments. Two new aspects that relate specifically to transgenic fish will be briefly commented upon. First, the hybrids produced by crosses between transgenic and wild fish will be heterozygous for the transgenic trait. The strength of natural selection against (or for) the new trait will depend on the expression of this trait in heterozygotes relative to homozygotes. A well established single-locus theory can be used to predict the fate of the transgenes at various levels of gene flow from transgenic to wild fish, and at various selection regimes and dominance relationships. It should be noted that when immigration rates into natural populations are very high, the recipient populations may be swamped by the inflowing genes irrespective of the strength of selection (Haldane 1932). It should also be noted that in small populations, stochastic processes could well override deterministic ones, so that the deterministic theory referred to here would be a very naïve description of genetic change.

Second, transgenic fish in aquacultural production will have gone through one bottleneck more than traditionally bred fish (Kapuscinski and Hallerman 1991). This results from the inbreeding which occurs when homozygous lines are produced from experimentally established transgenic individuals. Escapes of transgenic fish can therefore lead to an even more rapid loss of genetic variation in the recipient native populations than escapes of cultured fish, other factors being equal (Ryman 1991).

When is there too much interbreeding? Ryman (1991) suggested to use the level of naturally occurring gene flow as a quantitative guideline for permissible introgression rates into wild fish populations. Based on the relationship between observable genetic differentiation among populations (G_{ST}) and the number of effective migrants per generation in Sewall Wright's island model, the acceptable introgression rate for an average G_{ST} of 10% (a common figure for salmon populations) would be only 2 individuals per generation. This is a long way off the massive numbers of escaped cultured salmon now spawning in Norwegian rivers, and should mandate an entirely new approach to containment of cultured salmonids.

Altered use of cultured fish

In my opinion, the possible area expansion of aquaculture following production of transgenic cold-tolerant salmon could lead to the most dramatic environmental effects of the production of transgenic fish. This contention is based on (1) the regrettable fact that the current administration of Norwegian aquaculture does not appear to take the threats posed by escapes seriously, and on (2) the observation that dramatic biological effects on native populations can occur whether the escaped fish spawn successfully or not.

The natural populations in the northernmost parts of the distribution area of Atlantic salmon on both the American and Eurasian continent have been spared from the effects of aquaculture, simply because aquaculture is not there. It is likely, however, that the use of transgenic fish expressing antifreeze proteins would remove the major constraint on expansion of aquaculture to the north. This would mean that an increasingly smaller number of natural populations would remain reasonably unaffected by aquaculture, and could easily lead to loss of unique genetic resources. One major area for concern, therefore, would be the genetic and ecological integrity of local populations of Atlantic salmon in northern rivers. Some of these rivers, like the Tana in Norway and the Petchora in Russia, are among the most important rivers for natural production of Atlantic salmon today.

Another area for concern would be the effects on Arctic charr populations which thrive in northern streams where other salmonids are absent or rare from temperature limitations. For example, the islands of Svalbard have river systems in areas that were deglaciated 100,000 years ago, ten times as long ago as mainland Norway. The Arctic charr populations in these streams - and the whole stream ecosystem - represent irreplaceable biological resources which could easily be destroyed by expansion of aquaculture to the north.

Beyond salmonids

The perspective of this report has hitherto been that of interactions between transgenic and wild salmonids in coastal streams. However, the effects of salmonid fishes -including transgenic ones- on other trophic levels need also to be considered. Recent experimental evidence suggests that salmonid fishes have strong effects on the community dynamics of several trophic levels in river food webs; predatory and herbivorous insects, and macro- and epiphytic algae were all affected by release of rainbow trout (Power 1990). Transgenic fish with enhanced growth rate and larger appetite would have different effects on the stream ecosystem than native fish of the same species. First of all their larger size would make them more competitive versus other fishes, and they would be able to consume larger-sized prey that were not available to non-transgenic individuals.

Any use of genetically modified organisms must be viewed against the background of introductions of living organisms being a major cause for loss of global biodiversity. It has for example been estimated that introductions have contributed to

68% of the extinctions of North American fish species during the past century. The effects of introductions on other trophic levels have seldom been documented, although well known groups like freshwater mussels are known to be highly vulnerable to introduced species (Allan and Flecker 1993). If aquaculture of transgenic Atlantic salmon led to a northwards expansion of this species, this would imply introduction of a keystone predator to freshwater invertebrate communities, and could lead to a cascade of extinctions in northern fresh waters.

SCENARIOS FOR TRANSGENIC FISH IN AQUACULTURE

In the following, I present three different scenarios for a salmon aquaculture based on transgenic fish, and examine to what extent they are compatible with established goals for biological conservation. The goals for conservation can be stated as "maintaining as much as possible of the biological diversity at all levels from genes to ecosystems".

Scenario 1: Profitability

The first scenario assumes that transgenic fish will be used in accordance with their profitability, and represents an extrapolation of current trends in salmon aquaculture. This will inevitably lead to more populations being lost because of disease introduction and replacement by escaped fish. Local adaptations will continue to be eroded because of interbreeding between escaped and wild fish. Moreover, no readaptation to local environments will take place as long as the escapes continue. Expansion of aquaculture into new areas following the use of cold-tolerant transgenic fish can only make this situation worse because more natural populations of salmonids, and a variety of other species, would be at risk. Of course, such a development is not compatible with goals for biological conservation.

Scenario 2: Physical containment

An alternative scenario is one based on development of fish farms where physical containment is much better than in net pens, and where in- and outflowing water can be controlled for pathogens and waste products. Such fish farms have been constructed for land-based operations, and they have some obvious positive effects on fish farming itself such as the economic benefit of not raising fish that escape. Concerns about the environmental effects of transgenic organisms could well serve as an incentive for better physical containment in aquaculture. But is this good enough? Physical containment measures are subject to disruption, for example by extreme weather conditions or by human error. A single escape leading to establishment of a species beyond its range or to large-scale hybridization between cultured and wild fish can have negative long-term ecological and genetic effects. Therefore, to rely only on physical containment cannot be considered consistent with goals for biological conservation.

Scenario 3: Biological and physical containment

The third scenario is based on the suggestion that gene technology be used to drastically decrease the fitness of cultured fish, should they escape to the environment. Transgenic salmonids constitute a 'high-risk' group according to international checklists for evaluating the safety of releases of transgenic organisms (e.g. Tiedje et al. 1989). If they are to be used in large-scale production, one of the goals for genetic engineering should be to cripple those fish that may eventually escape. Salmonids can be sterilized on a large scale by technologically simple genetic manipulation. Such sterile fish would not have direct genetic effects on natural populations, but could have indirect genetic effects through the reduction of population sizes.

It has been suggested that genetic modification be used to develop suicide gene constructs that would not only prevent the fish from reproducing but would actually become lethal at an appropriate time. For example, Maclean and Penman (1990) suggested to develop nutritionally crippled strains of fish which would require a particular amino acid that were not present in the natural diet of salmon. When and how should such a gene be expressed? Consider the following scenario (which was realized in Norway in autumn of 1991, except no transgenic fish was involved): A major escape occurs just prior to or within the spawning season. If they had not been sterilized, the fish that escaped would not eat and would therefore not be affected by any nutritional crippling. Instead, they would enter the nearest stream to spawn and would pass the lethal gene to the next generation. Moreover, the vast majority of matings in that stream would involve transgenic fish as one parent, and entire cohorts could be lost if the lethal gene was dominant. If they had been sterilized, they would nevertheless be able to interfere with the spawning of the native fish if they entered the stream, but could have minimal effects if they migrated to the open ocean. Thus, use of genetic modification to cripple cultured fish must be carefully considered to incorporate all the possible biological situations encountered. But such studies should definitely go along with studies that use gene technology to improve production characteristics. Only when environmental safety is one objective of the experimentation with transgenic fish can production goals be combined with internationally established goals for biological conservation.

CONCLUSION

Experience from releases of non-native populations of salmonid fishes suggests that the genetic effects on natural populations are negative. Likewise, the ecological effects caused by introductions of fish species are negative to native biota, and introductions are frequently registered as a cause of species extinctions. Current aquaculture involves rearing techniques which often result in escape of cultured fish to natural environments. This requires that any use of transgenic fish in aquaculture must be evaluated as a release experiment. The use of transgenic Atlantic salmon in aquaculture has a potential to increase the range of effects on natural environments. In particular, the use of transgenic salmon expressing antifreeze protein genes could lead to a northwards expansion of aquaculture, which would put northern populations of Atlantic salmon, Arctic charr and a number of invertebrates - all of them are so far spared from the effects of aquaculture - at risk of irreversible change or even extinction. Transgenic fish should not be used in aquaculture as long as the aquaculture industry is so poorly adapted to coexistence with wild populations. It is suggested that gene technology be used in an active search for biological containment of cultured fish in order to develop aquaculture along lines which are compatible with goals for conservation of biological diversity.

REFERENCES

Allan, J.D. and Flecker, A.S. (1993): Biodiversity conservation in running waters. BioScience, 43:32-43.

Chourrout, D. (this volume):

Haldane, J.B.S. (1932): The causes of evolution. Longmans, Green & Co., London. (Reprinted 1990 by Princeton University Press, Princeton)

Hindar, K. (1993): Genetically engineered fish and their possible environmental impact. NINA Oppdragsmelding (Trondheim), 215:1-48.

Hindar, K., Ryman, N. and Utter, F. (1991): Genetic effects of cultured fish on natural fish populations. Can. J. Fish. Aquat. Sci., 48:945-957.

Kapuscinski, A.R. and Hallerman, E.M. (1991): Implications of introduction of transgenic fish into natural ecosystems. Can. J. Fish. Aquat. Sci., 48 (Suppl. 1):99-107.

Krueger, C.C. and May, B. (1991): Ecological and genetic effects of salmonid introductions in North America. Can. J. Fish. Aquat. Sci., 48 (Suppl. 1):66-77.

Maclean, N. and Penman, D. (1990): The application of gene manipulation to aquaculture. Aquaculture, 85:1-20.

Power, G. (1969): The salmon of Ungava Bay. Arctic Inst. N. Am. Tech. Pap., 22:1-72.

Power, M.E. (1990): Effects of fish in river food webs. Science (Wash.), 250:811-814.

Powers, D.A., Chen, T.T. and Dunham, R.A. (1992): Transgenic fish. In Murray, J.A.H., ed. Transgenesis: applications of gene transfer. Wiley, Chichester, pp. 233-249.

Ryman, N. (1991): Conservation genetics considerations in fishery management. J. Fish Biol., 39 (Suppl. A):211-224.

Taylor, E.B. (1991): A review of local adaptation in Salmonidae, with particular reference to Pacific and Atlantic salmon. Aquaculture, 98:185-207.

Tiedje, J.M., Colwell, R.K., Grossman, Y.L, Hodson, R.E., Lenski, R.E., Mack, R.N. and Regal, P.J. (1989): The planned introduction of genetically engineered organisms: ecological considerations and recommendations. Ecology, 70:298-315.

Williamson, M., Perrins, J. and Fitter, A. (1990): Releasing genetically engineered plants: present proposals and possible hazards. Trends Ecol. Evol., 5:417-419.

PLENARY DISCUSSION AT THE END OF THE SESSION

Question:
I would like to thank dr. Hindar for having the courage to look at the future and at what might be the long term ecological effects of these biotechnological applications to aquaculture. I would appreciate if dr. Chourrout could respond to that.

Response:
He is not for nor against these applications. The question is to know what will be the pressure to make transgenic fish. My feeling is that the pressure to increase growth will be quite weak but the transfer of antifreeze protein gene into fish might be quite important for the East Coast of North America. The problem in Norway is due to the overproduction of salmon at a very low price. That is not the case for Atlantic salmon in North America. This idea of area expansion is important. In the case of enhancing growth by selective breeding the escape of such fish into the wild is also an important issue. But I doubt whether there is another solution to cold tolerance than to transfer antifreeze protein genes. I myself will not push that but there certainly is a pressure to do so.

Question:
I think that dr. Hindar's arguments are quite logic. Selective advantage could be argued and his next conclusion is try to apply some form of biological containment, be it suicide genes or whatever. What are your views on the possibilities for that?

Response:
That is probably an illusion.

Question:
What is the tolerance of the antifreeze fish and what are the water temperatures that have been talked of there?

Response:
In the Arctic seawater temperatures may go as low as minus 2.5 degrees centigrade. Atlantic salmon survives temperatures down minus 0.5 degrees centigrade. If antifreeze protein genes are inserted into Atlantic salmon or rainbow trout, they survive colder seawater temperatures. As dr. Chourrout mentioned, the transgenic Atlantic salmon expressing antifreeze protein genes do not yet express it at levels that would give them resistance to so cold seawater.

Question:
I was wondering why there not better efforts being made to physically contain the fishes in the fish farms? It seem that they are being grown unseparated from the environment. Would it not be sensible to increase containment a little bit? That might be rather helpful.

Response:
Economy!

Question:
Do you want wild salmon in Norway or not?

Response:
I can give easy answers for myself and I can give more complicated answer for Norway. As I see it, the salmon disaster is caused by the fact that it is managed by three different departments. Wild fish are being administered by the Department of Environment, cultured fish by the Department of Fisheries and fish that has some disease by the Department of Agriculture whether or not they are wild or cultured. The Department of Environment is certainly concerned about the effects of farmed fish on wild populations but, as I see it today, wild fish are in reality managed by the Department of Fisheries as long as they let so many fish escape as happens today.

Question:
I don't think we should let the moment pass without pointing out that the ecological impacts of this kind of work of genuine range extension with an organism that could well be a keystone species. It could constitute problems of several orders of magnitude greater than any of the problems that we have talked about so far. These are the kinds of concerns that really would get ecologists extremely agitated, whereas the others are things about which the ecologists would be sanguine.

Response:
I agree and I should emphasize that I have restricted my discussion to the effects of salmon on other salmonids and that I have not all talked about effects on freshwater ecosystems.

CONCLUDING REMARKS ON FISH

Dr. C. Pla (rapporteur), Laboratory of Ichthyology (Genetics), University of Girona, Spain

1. Long term is between 10 and 100 years. In addition, long term should also be determined by a time period during which regulators and risk-assessors have the possibility to exert influence on predicting the potential ecological impacts of transgenic fish, in order to avoid negative effects.

2. Ecological effects are displacement of wild populations and introgression.

3. Unwanted effects can be eliminated through the use of sterile fish and confinement.

4. The risk assessment methodology for the short term consequences can also be applied for the middle term effects but it is difficult to predict for the long run.

5. There are crucial differences between terrestrial and aquatic environments. Fluvial ecosystems (rivers) know more variability than terrestrial ecosystems; it is possible to find different ecosystems in a river along its course. Marine ecosystems are more stable and open than terrestrial ecosystems.

6. Transgenesis could increase the genetic variability of a modified stock from its status and, initially, this is good news for any population. However, if this stock is released in an ecosystem where there is a wild stock, the genetic divergence between both stocks may increase, despite that both stocks initially had the same origin. The result may be competition between the two populations and the displacement of either one of those two populations.

THE POTENTIAL ROLE OF MONITORING IN RISK ASSESSMENT AND IN FUNDAMENTAL ECOLOGICAL RESEARCH

INTRODUCING THE POTENTIAL ROLE OF MONITORING IN RISK ASSESSMENT AND IN FUNDAMENTAL ECOLOGICAL RESEARCH

Dr. R. Harrington (general rapporteur), Conservation Genetics and Mammal Research, National Parks and Wildlife Service, Ireland

Biological monitoring implies that the fate of an organism, a population or a community is observed through time. With the emphasis of this conference on the potential long term ecological impact of GMOs, monitoring the fate of GMOs following their deliberate release into the uncontained environment is central to the discussion.

Current monitoring procedures are short term and applied to determining whether or not the GMO or its acquired genetic component, has been contained within the release site. Present monitoring procedures cannot be considered as addressing the ecological and long term impact of GMOs on the natural environment. While present monitoring procedures may be adequate to address the current scale of the uncontained use of GMOs, the prospect of ever increasing scales in the use of GMOs challenges biotechnologists and ecologists to better anticipate the gene flow and ecological impact of GMOs upon release.

In this section Dr. Chris Gliddon will examine the information that is presently available to assess GMO impact on the natural environment. He will present the theoretical background and how experimental design influences isolating distances in monitoring. He will emphasize the need to understand what is presently predictable and what is not, through the presentation of case studies.

The presentation will state the need for fundamental ecological research and that long term usually means longer than one human life - or career, when applied to biological interactions.

THE POTENTIAL ROLE OF MONITORING IN RISK ASSESSMENTAND IN FUNDAMENTAL ECOLOGICAL RESEARCH

Dr. C. Gliddon, School of Biological Sciences, University of Wales, United Kingdom

SUMMARY

In the context of the release of genetically modified organisms (GMOs), monitoring should be designed to aid in the prediction of potential environmental impact or risk. According to the classical risk assessment paradigm, risk is the result of the frequency of occurrence of an effect (the *exposure*) multiplied by the magnitude of that effect (the *hazard*). For a full assessment of such an environmental impact or risk, monitoring should, therefore, be designed to quantify both the exposure and the hazard.

The present protocols for monitoring of the release of GMOs focus, almost exclusively, on detection of possible escapes from trials and the auditing of the efficacy of the biological containment of those trials. At best, such monitoring only provides some information relating to the possible levels of exposure but, since containment measures are explicitly designed to preclude escape, the generation of information relating to the quantification of hazard is prevented.

The design of monitoring protocols to measure effectively exposure and hazard will be discussed in the context of three case studies. First, the case of gene transfer from wild beet to cultivated beet will be discussed with particular attention to the likely consequences of the introduction of commercial varieties of sugar beet, which have been genetically modified to carry a gene for resistance to a herbicide. Second, the role of hybridization between introduced and indigenous species, followed by repeated backcrossing of the hybrid to the wild species will be illustrated using the example of *Senecio squalidus* and *S. vulgaris* in Britain, with particular reference to the lessons for monitoring procedures. Finally, the utility of a classical population biological approach to risk assessment will be shown, using work on the population structure of wild or feral populations of *Brassica oleracea* ssp. *oleracea* and *Beta vulgaris* ssp. *maritima*.

In summary, problems associated with direct monitoring of releases of GMOs will be discussed and the suggestion made that experiments which are normally regarded as being of a fundamental ecological nature may be far more relevant to risk assessment of GMOs than the currently fashionable 'near-market' approach.

INTRODUCTION

This point in time represents a transition between the confined use in the environment of genetically modified organisms, primarily in small-scale field trials for the purpose of research and development, and a scaling-up of releases prior to moving on to their unconfined commercial release. While there is still no harmonization of regulations relating to the release of GMOs on a global scale, there is a large degree of conformity in the aims of the regulations, namely prevention or minimization of environmental damage. To this end, the assessment of risk associated with the release of GMOs into the environment is required during the research and development stage with the intention that salient information obtained from monitoring those small and medium scale releases will be used to improve the assessment of risk associated with potential commercial release. The present protocols for monitoring of the release of GMOs focus, almost exclusively, on detection of possible escapes from trials and the auditing of the efficacy of the biological containment of those trials. At best, such monitoring only provides information relating to the possible frequency at which escapes may occur and, since containment measures are explicitly designed to preclude escape, the generation of information relating to the quantification of likely environmental impact is prevented. There is a clear need to define the information necessary to permit appropriate risk assessments to be made, thereby allowing an informed view to be made of the likely long term ecological effects of the release of GMOs into the environment.

Concepts in Risk Assessment

Mackenzie and Henry (1990) described the present paradigm in evaluating risk as follows:

$$risk = exposure \times hazard$$

In the application of this conceptual relationship to GMOs released into the environment, *exposure* is a measureof the organism's (insert's) ability to escape from its place of deliberate introduction. The subsequent fate of any escaped organisms (inserts) then needs to be quantified in terms of the likelihood of persistence, increase and spread in the environment.

The third term in the equation, *hazard*, refers to the impact of the escaped organism on the existing ecosystem.

In some senses, this definition of risk is somewhat different from that which has been applied, for example, to the nuclear industry. In the application to GMOs, *exposure* includes both the probability of an escape occurring and the likelihood of persistence, increase and spread in the environment. That is, *exposure* is not simply a probability but an expected number, which includes effects of the magnitude (scale) of the escape as well as the subsequent fate of the escaped organisms or genes.

In order for monitoring information to be useful in addressing the problem of quantification of *exposure* for use in risk assessment, it is necessary that it produces data allowing the estimation of the probability of escape as a function of distance from the intentional introduction, together with some estimate of the effect of size of the introduction and some measure of the likelihood of persistence.

The quantification of *hazard* or impact of the escaped organism on the existing ecosystem has rarely been addressed by monitoring of field trials, not least because field trials have been designed primarily to preclude the possibility of escape and, hence, of any *hazard* occurring. Furthermore, the quantification of *hazard* should involve both biological and socio-economic criteria since, for example, the local extinction of a species certainly has an impact on the local ecosystem but the question of whether this extinction constitutes damage rather than benign change requires a value judgment in its answer.

Existing Monitoring Protocols

At the present time, prior to any releases being carried out in Member States of the EC, a risk assessment must be carried out ('Deliberate Release Directive', 90/220). In general terms, this assessment comprises an examination of the potential routes of escape of the GMO or inserted construct coupled with speculation concerning any possible environmental impact should such an escape occur. In the main, this information will have been gleaned from the scientific literature together with experience gained from monitoring previous releases of similar GMOs. The Competent Authority of the Member State in which such releases are carried out frequently requires that the release is 'monitored'. This 'monitoring' is essentially designed to ensure that any containment measures required by the 'Consent to Release' are efficacious. That is, such 'monitoring' is, in reality, an audit of the success of the containment measures. By definition, such an audit cannot provide any information of the magnitude of any hazard posed by the GMO since it is designed to prevent the hazard from occurring. At best, some information may be provided relating to the degree of exposure at varying (short) distances from the release site. The quantification of risk remains almost as difficult after such a release has taken place as it was prior to the release.

USEFUL BIOLOGICAL MODELS

Sugar-beet and Wild-beet

Recent studies, using modern molecular techniques, have been applied to looking at the relationship between the three forms of beet, *Beta vulgaris* (Boudry et al., 1992, 1993). The three forms are inter-fertile and are:

-	*Beta vulgaris* ssp. *vulgaris*	sugar-beet, fodder beet & beetroot
-	*Beta vulgaris* ssp. *maritima*	sea beet or wild beet
-	*Beta vulgaris* ssp. unknown	weed beet or ruderal beet

Weed beet was first reported as a problem to agriculture in the USA during the 1920's, with the situation worsening rapidly in California in the 1950's. The problem in northern Europe became apparent in the 1970's and is a serious problem for some sugar-beet growers in continental Europe and, to a lesser extent, the UK. The major difference between weed- and sugar-beet is the requirement for plant vernalization to induce flowering in the crop. This character is under a simple one-locus, two-allele control with the allele for bolting (B) dominant to that for non-bolting (b). The maritime populations of sea-beet in Europe show great variability for vernalization requirement. However, in the regions of sugar-beet seed multiplication in France, ruderal beet is mainly homozygous for the bolting allele. Recurrent hybridization between seed multiplication lines and ruderal beet has resulted in the transfer of the bolting characteristic to individuals within the sugar-beet varieties, with the resulting "weed-beet" problems when the crop is grown in northern Europe. An analysis of cytoplasmic markers in the sugar-beet varieties and weed-beet suggests that weed beet is the result of a cross between ruderal beet as the male parent and the crop as the female parent. Studies of ruderal populations around the peripheries of the main areas of sugar-beet seed production in France show that both nuclear genes and, less frequently, cytoplasmic genes from the crop have introgressed into wild or weed populations. While the environmental impact of this exchange of genes is not clear, there can be a clear lesson that could be applied to the likely consequences of the introduction of commercial varieties of sugar-beet, which have been genetically modified to be resistant to systemic, total herbicides. It is almost inevitable that, following seed multiplication of such varieties, introgression of pollen from adventitious ruderal-beet will result in herbicide resistant weed-beet problems in the sugar-beet growing areas of northern France, thus exacerbating an already serious weed control problem.

This information, which is clearly of relevance to risk assessment, has been obtained through the application of standard population biological techniques to an existing crop-weed relationship and has been far more informative, from the perspective of risk assessment, than any formally required 'monitoring' protocols could have been.

Figure 1: Pathways for gene flow in Beta vulgaris
(from Boudry et al., 1993)

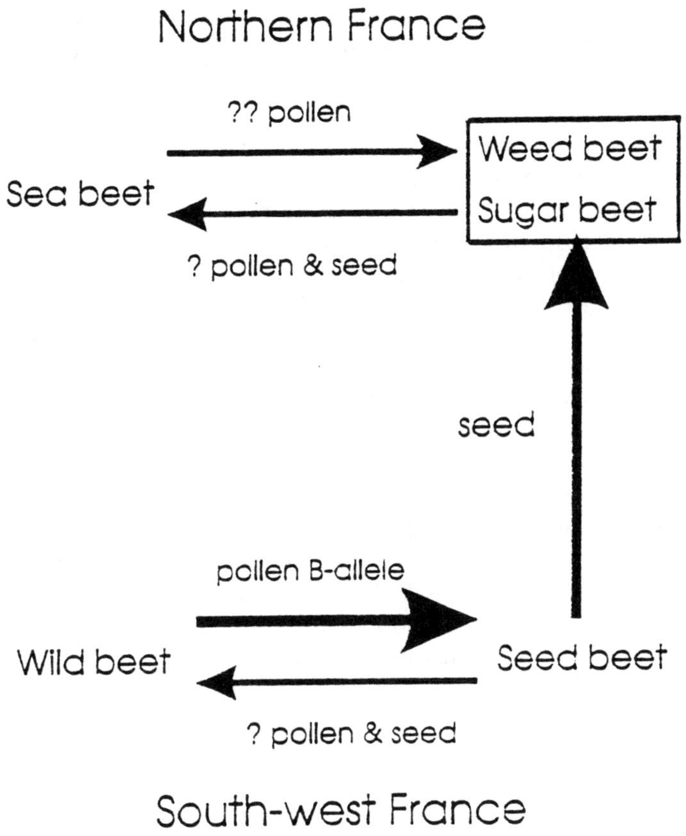

Senecio vulgaris* and *S. squalidus

Interspecific hybridization is one of the major mechanisms for generating new taxa in the plant kingdom (Stebbins, 1959, Stace, 1975, Grant, 1981). There are many cases in which hybridization between an indigenous and alien plant species has resulted in a new, sexually fertile taxon. One well-documented example of this is the result of a hybridization event between *Senecio vulgaris*, common groundsel, and *S. squalidus*, the Oxford ragwort, (see Abbott, 1992, for a review). *S. vulgaris* is a native herb of the British Isles. The common form, *S. vulgaris* var. *vulgaris*, produces a capitulum containing only disc florets, which tend to be highly self-fertilizing and to produce relatively non-dispersed seed. A more recent form, *S. vulgaris* var. *hibernicus*, differs from the common form in that ray florets are produced in the capitulum. Seed obtained from ray florets tends to be more outcrossed and to disperse further than that of disc florets. The difference in capitulum type is due to variation at a single locus (Abbott et al., 1992).

The first report of wild radiate groundsel was in 1832 from Oxford, England, following the escape from the Oxford Botanic Garden in 1794 of the radiate species *S. squalidus*, a native of Sicily. Since this time, *S. squalidus* has spread widely throughout the British Isles, often following railway tracks, whose limestone ballast provides a well-drained medium somewhat reminiscent of its native Sicily. At the present time, *S. squalidus* is often found growing alongside *S. vulgaris* in open, disturbed habitats. Highly self-sterile hybrids (*S.* x *baxteri*) are formed at low frequency with *S. vulgaris* as the female parent but backcrosses to *S. vulgaris* are fertile. The hybrid radiate groundsel, *S. vulgaris* var. *hibernicus*, is postulated to be a stabilized introgressant, resulting from repeated backcrossing of *S.* x *baxteri* with *S. vulgaris* var. *vulgaris*, on the basis of three lines of evidence: (1) the strong similarity of the pattern of spread of *S. squalidus* and *S. vulgaris* var. *hibernicus* throughout the British Isles over the past 160 years; (2) tetraploid plants, produced by backcrossing *S.* x *baxteri* to *S. vulgaris* var. *vulgaris*, bear a close resemblance to *S. vulgaris* var. *hibernicus*; and (3) an allozyme marker, absent from non-radiate groundsel populations but present at high frequency in *S. squalidus*, is found at intermediate frequencies in radiate groundsel.

Another species appears to have resulted from the original self-sterile hybrid, *S.* x *baxteri*, namely *S. cambrensis*. In this species, the original sterility barrier has been overcome by a doubling in ploidy from the hybrid (2n=30) to *S. cambrensis* (2n=60). Evidence from isozymes and cpDNA (Ashton & Abbott, 1992, Harris & Ingram, 1992) has confirmed the allopolyploid nature of *S. cambrensis* and shown it to have originated in at least two separate localities.

Figure 2: Pathways for gene flow in Senecio vulgaris and S. squalidus (redrawn after Abbott, 1992)

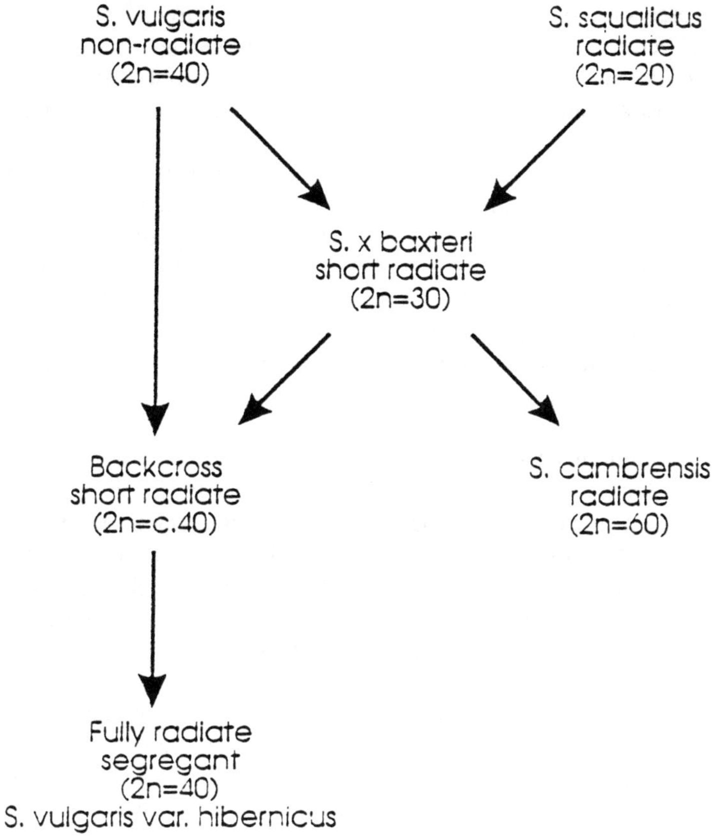

The example above is one of several indicating that self-sterility of a hybrid is not a sufficient barrier, *per se*, to invasion by genes derived from an alien into an indigenous species. Indeed, the production of fertile allopolyploids and stabilized introgressants are well documented, and not uncommon, phenomena whereby genes from exotic species have been combined with those of the native flora. If *S. squalidus* had been a GMO, a hypothetical risk assessment of the scenario described above would probably have decided that, since the hybrid was 'sterile', the exposure would be very low and therefore the risk of escape would be minimal!

PREDICTION OF RATE OF SPREAD OF INTRODUCED GMOs

In order to quantify the rate of increase and rate of spread of any gene or organism in a population, two basic types of information are required: an estimate of 'fitness', at least in terms of survivorship and fecundity; an estimate of the magnitude of gene-flow or dispersal in existing natural populations into which the gene or organism may escape.

The measurement of fitness is notoriously difficult. This is primarily because fitness is environment dependent and the environment includes both organisms of the same and other species, together with the genetic background of the organism into which a gene has 'escaped'. Where the result of an escape is a hybrid between a crop and a related indigenous species, the problem of appropriate fitness measurement is compounded by the need to measure the fitness of not only the F_1-hybrid, but also that of backcross generations (see the example of *Senecio* species above).

The rate of spread of an organism is determined in part by fitness but, in cases where fitness differentials are slight, spread is mainly determined by the magnitude of gene-flow or dispersal and the connectivity of the recipient natural populations. Gene-flow and dispersal can be measured by direct experimentation. However, the scale of such experiments is often constrained by time and cost to be much smaller than is desirable. In particular, there is a need to measure events that are rare at small scale but which will become frequent events as the scale of release of GMOs increases.

An alternative to the estimation of spread by direct experimentation is to apply methods developed in population genetics theory to indirectly estimate rates of gene-flow in natural populations (Goudet et al., 1994). Such indirect methods are problematic in that they need natural populations. However, in cases where the perceived risk involves spread of escaped genes through populations of crop relatives, these methods are ideal since they combine the effects on rate of spread of both gene-flow and population structure (i.e. connectivity).

The application of these indirect methods has been made using natural populations of wild cabbage, *Brassica oleracea* ssp. *oleracea* and sea-beet, *Beta vulgaris* ssp. *maritima*. These methods have allowed estimates of the magnitude of gene-flow (including population connectivity) to be made up to 8 km distant from a source.

The shape of the dispersal curves allows an accurate estimate to be made of the frequency with which genes will spread from an original introduction. It is interesting to compare the estimates of dispersal obtained from these indirect methods for beet, with those obtained by direct experimentation on pollen-flow (e.g. Darmency and Renard, 1992). It is apparent that the direct experimental methods have dramatically underestimated the magnitude of long distance dispersal, thereby significantly underestimating probable exposure levels.

Figure 3: Gene flow in Sea Beet

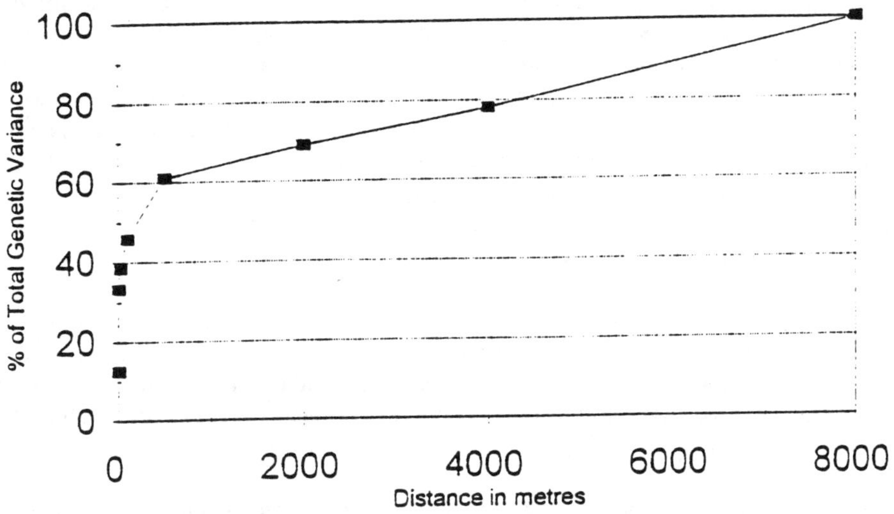

Figure 4: Gene flow in Wild Cabbage

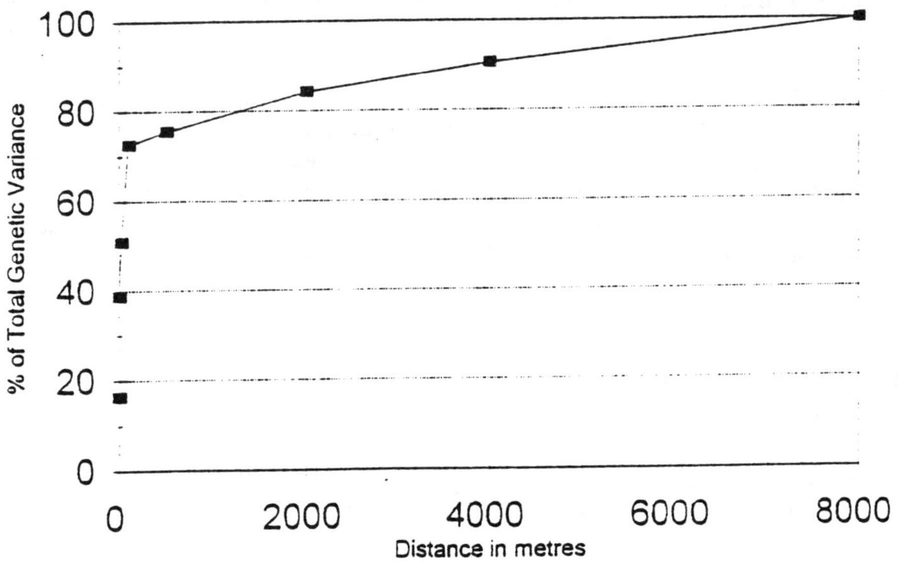

CONCLUSIONS AND RECOMMENDATIONS

The first step in any risk assessment must be the identification of genetically modified organisms that have the potential to do environmental damage. In the context of crop plants, this will generally entail identifying those species with weedy, or potentially weedy, wild or feral relatives since it is for this class of GM crops from which genes will escape and that, therefore, may pose a risk.

The second step, for those GMOs where exposure is likely to be non-zero, is to address questions relating to the fitness of the GMO itself and also of the hybrids and backcrosses that will be generated by the escape of genes. It has been pointed out that fitness estimation is notoriously difficult. Many traits affecting fitness are only expressed under certain environmental conditions (e.g. high density, see Reboud and Till, 1991) and the effects of many genes on the phenotype of an organism are strongly dependent on their genetic background. In many instances, it appears highly unlikely that sufficient relevant information could be accrued by direct experimentation and, under those circumstances, it would appear preferable to err on the side of caution when considering applications for permission to release GMOs.

A third step is to estimate the rates of spread of genes through wild and feral populations of related species. The information provided by this approach will be of use in identifying precise levels of exposure and, in addition, generate the type of information needed to design 'mitigation procedures' should an unforeseen environmental impact occur.

Current 'monitoring practices', attached as conditions to the permission to release a GMO, seldom, if ever, improve the information relevant to risk assessment. This problem is particularly apparent if hazard is to be quantified since the audit procedure of the release is usually designed to prevent escapes from occurring and, hence, a hazard from existing.

There is clearly a need for more fundamental ecological research to be carried out in the areas described above, to permit more reasoned assessment of the risk associated with the release of GMOs into the environment. It is incumbent upon the scientific community to ensure that adequate funding exists to allow this research to take place and to realize that a great deal of pertinent information can be obtained from the use of biological model systems.

REFERENCES

Abbott, R.J. (1992) Plant Invasions, Interspecific Hybridization and the Evolution of New Plant Taxa. Trends in Ecology and Evolution, 7, 401-405.

Abbott, R.J., Ashton, P.A. and Forbes, D.G. (1992) Introgressive origin of the radiate groundsel, *Senecio vulgaris* L. var. *hibernicus* Syme: Aat-3 evidence. Heredity, 68, 425-435.

Ashton, P.A. and Abbott, R.J. (1992) Multiple origins and genetic diversity in the newly arisen allopolyploid species, *Senecio cambrensis* Rosser (Compositae). Heredity, 68, 25-32.

Boudry, P., Saumitou-Laprade, P., Vernet, Ph. and Van Dijk, H. (1992) Les betteraves mauvaises herbes: origine et évolution. In: IXème Colloque Internationale sur la Biologie des Mauvaises Herbes. Editions INRA.

Boudry, P., Mörchen, P., Saumitou-Laprade, P., Vernet, Ph. and Van Dijk, H. (1993) The origin and evolution of weed beets: consequences for the release of herbicide resistant transgenic sugar beets. Theoret. Appl. Genet. (in press).

Darmency, H. and Renard, M. (1992) Efficiency of safety procedures in experiments with transgenic oilseed rape. In: Casper, A. & Landsmann, J. (Eds.). Proceedings of the 2nd International Symposium on the Biosafety Results of Field Tests of Genetically Modified Plants and Microorganisms. Biologische Bundesanstalt für Land- und Forstwirtschaft, Braunschweig, Germany. pp. 54-60.

Goudet, J., De Meeus, T., Day, A. and Gliddon, C. (1994) The different levels of population structuring of dogwhelks, Nucella lapillus, along the south devon coast. In beaumont, A. (Ed). Genetics and Evolution of Aquatic Organisms. (in press).

Grant, V. (1981) Plant Speciation (2nd. edn.). Columbia University Press.

Harris, S.A. and Ingram, R. (1992) Molecular systematics of the genus *Senecio* L. I. Hybridization in a British polyploid complex. Heredity, 69, 1-10.

MacKenzie, D.R. and henry, S.C. (1990) Towards a consensus. In: MacKenzie D.R. & Henry S.C. (Eds.). Biological Monitoring of Genetically Engineered Plants and Microbes, Proceedings of the Kiawah Island Conference. Agriculture Research Institute, Bethesda, Maryland, USA, 273-283.

Reboud, X. and Till, I. (1991) The cost of herbicide resistance measured by a competition experiment. Theor. Appl. Genet., 82, 690-696.

Stace, C.A. (1975) Hybridization and the Flora of the British Isles. Academic Press.

Stebbins, G.L. (1959) The role of hybridization in evolution. Proceedings of the American Philosophical Society, 103, 231-251.

PLENARY DISCUSSION

Question:
Towards the end you said something very curious: Fitness was not worth measuring because these populations are going to be ephemeral.

Response:
I think I said genotypes are going to be ephemeral.

Question:
Anyway, in general terms: In the absence of knowledge on fitness, how can you make an informed risk assessment?

Response:
You need to speculate in an informed manner. What I said was that I think that experiments to measure fitness will not in the main improve the level of informedness of your speculation because one tends to measure the wrong things or things, if they are genotypes, which are only ephemeral; ephemeral in the sense that you are not going to retain very many hybrids for very long; back cross 1, back cross 2, back cross 3, and as time goes by you will get an increasing jumbling up of all of the genotypes. Well, one needs to have some idea about the fitness and clearly if you have something which you think will produce an incredibly aggressive weed, you worry it will get out. If someone were crazy enough to want to let it out, you would have to do a very large amount of experimentation. But I doubt very much, giving how wonderfully socially aware our industry is, that they would ever want to do such a thing. So, I don't think that is a question we will have to face. It is certainly a question we might need to face, if a crazy academic from a university wanted to play with GMOs because they tend to be not socially aware.

Question:
I was impressed by your graphs on gene flow but how long has the flow lasted, since the longer the flow lasted, the further a gene might go?

Response:
That is certainly the case but basically what you are doing is looking at the percentage of the genetic variance for which you can account. Now, you need to have allowed the particular gene flow patterns to have existed for long enough for you to generate a reasonable dynamic equilibrium in those populations. What you simply can read those as being is not the absolute distances of gene flow but quite simply saying: If there is a certain amount of gene flow from the point source at what rate does it decay as you go away. And basically, if you do this sort of diffusion experiments by computer, you find that you reach a sort of dynamic gene flow pattern. But clearly, if one waited for long enough, any one gene can go to any one other place providing they have connections.

Question:
First, I regret that we have so little time to discuss such an important topic. I have to criticize the organizers for having scheduled not enough time since the monitoring issue is of course relevant to all other sessions of this conference. I agree with most points of the speakers but I think you can only compare small events with GMOs to natural events. I think there is point you have to cross the river Rubicon where you have to work with GMOs because you use genes that you will never see naturally. If plants can get new genes from relatives, it is very important to work with natural systems but you must also examine the consequences of the new genes in the species and this work can only be done with transgenic organisms in my opinion. You have to examine the fitness of the transgenics but you should be aware what kind of genotypes you are testing.

Response:
Sure, for measuring fitness one must obviously use the GMO. I do think that in the main the sorts of GMOs that we are seeing and will be seeing are unlikely to be of a type that will dramatically increase the fitness of anything capable to escape. That is a statement of the situation now. I personally don't think that giving herbicide resistance to sea beet is likely to cause sea beet any problem at all. It might make it slightly less fit under which circumstances it won't spread much in the population. I don't think we are going to see any degradation in the natural environment. So, basically what I was wanting to do was to focus on estimating rates of spread for which it is completely unnecessary and even forcing bad experimental design in the main to measure spread using GMOs. It is far better to use nature which has been doing it for a long time and from which we can extract information. Now that has one plus side if you are a worrying type. Suppose something untoward happens, suppose my informed speculation about the likely impact of a release turns out to be misplaced and something bad has happened. If at least you know what the spread is and the rates of spread, then you can at least attempt to design mitigation exercises: you can attempt to put gene breaks in to stop further spread. So, from the level of caution you know at least how wide the fire breaks have to be.

SOME ECOLOGICAL MISCONCEPTIONS

Dr. M.J. Crawley, Centre for Population Biology, Imperial College, United Kingdom

MISCONCEPTION	**EXPLANATION**
THERE ARE STRESSFUL ENVIRONMENTS AND BENIGN ENVIRONMENTS	All environments are stressful for the organisms that inhabit them: the struggle for existence is ferocious everywhere; natural enemies want to eat your children; competitors want to move into your house.
YOU CAN "CHARACTERIZE THE ECOLOGICAL INTERACTIONS" THAT CAUSED A GIVEN EVENT	This is impossible without detailed experimentation. Suppose you introduce GMO species A and species B goes down: 1) interference (behavioural competition) 2) exploitation (resource competition) 3) allelopathy (chemical interference, toxins, etc.) 4) apparent competition (shared natural enemies)
IF SOMETHING GOES WRONG, WE CAN TIDY UP LATER	Release is forever; Problems are not recognized until they are irreversible.
GROUP SELECTION	Traits do not evolve "for the good of the species". In the Struggle for Existence, cheats prosper. Altruism can evolve through kin selection. We need to be aware of the GMOs inclusive fitness (i.e. of copies of the transgene in brothers, sisters and cousins).
WE SHOULD "GIVE ATTENTION TO POTENTIAL LONG-TERM ECOLOGICAL EFFECTS PRIMARILY THROUGH MONI-TORING"	You can't monitor potential effects, only actual effects; unless you know the actual effects to monitor, you don't know what to look for and so you certainly won't find it.

COMPETITIVE ABILITY IS A GENOTYPE-SPECIFIC TRAIT	Wrong! Competitive ability depends upon environmental circumstances, and upon the identity of the competitors, and upon densities of natural enemies. It is parameter values, not intrinsic properties that matter.
DEATH REDUCES POPULATION DENSITY	Not if there is sufficiently strong density dependence.
HIGHER REPRODUCTION MEANS HIGHER DENSITY	Not if there is sufficiently strong density dependence.
COMPLEX DYNAMICAL BEHAVIOUR REQUIRES COMPLEX MODELS TO DESCRIBE IT	Wrong! The 1-parameter time-lag difference equation can produce stable point equilibrium, stable limit cycles, or chaos.
DIFFERENTIAL PERFORMANCE OF PHENOTYPES IMPLIES EVOLUTION	Not unless there are genetic differences between the winners and losers.
COEVOLUTION IS THE RULE UNDER THE BALANCE OF NATURE	(e.g. pathogens evolve towards reduced virulence, hosts evolve towards immunity) Evidence suggests that reciprocal (true "co"evolution) is unusual. Evolution leads to intermediate virulence in pathogens, because cheats with higher virulence would prosper at the expense of less virulent forms.

PLENARY DISCUSSION

Question:
You seem to be putting the emphasis on the substrate which gives the parameters for expansion and for colonization and so forth. Would you like to elaborate upon the idea of just substrate and how we could anticipate. You ended your first talk with a slide of the west coast of Ireland and then saying: How unexpected would it be to find this moss running rampant. But in reality it is the perfect substrate because we have a beautiful moist metic site for mosses to grow. So, one could anticipate and indeed just around the corner where you took that photograph we have *Gunnera* which is six to twelve foot high in places and we have *Rhododendron ponticum* down at the south-west coast running right up along to the central-west coast; in fact it is also a problem in Scotland and in parts of Wales. So, really you could anticipate from the substrate-level where you would expect problems. Therefore, I would like perhaps to add to the predictive basis by looking at substrate that may be you could anticipate where problems might lie.

Response:
You are absolutely right about using alien plants that become extremely abundant. One does want to learn about the absolute maximum that one can from looking at those. If you do that, you can certainly rank the European habitats in order of their invasibility. But the surprising thing was that heathen and moorland rather than the kinds of woodlands that have been invaded by rhododendrons. That habitat on the slide is really very low in aliens before this moss got there and has very high cover of other native briars. So, it was not a obvious vacant niche for this moss. This moss has succeeded by a kind of outcompeting.

Question:
We are talking about resource-base upon which populations can build up their populations. The whole concept of where one might look at to anticipate what this resource-base might be. Do you think, as you just suggested, that you could zone Europe or indeed the world for problem areas for certain species?

Response:
I have many ambitions but zoning the world isn't one of them. I don't know; I think that is terribly difficult!

Question:
What fascinates me here is that there seems to be a strong difference between what is normally done today in risk assessment and what you said:

It is not the intrinsic properties of organism that matter. You place the emphasis far more upon the environment, whereas you look at today's risk assessment programmes and how they are carried out: There is this heavy emphasis on the host organism, on the vector used and on the insert used, and at the end of the story there are two or three lines on the environment. Can I conclude form what you said: It is the other way around? Should we forget about the characteristics of the organism? Are there no hazardous organisms to your opinion?; is it the environment that decides?

Response:
The last thing you want to do is to forget about anything and certainly not the attributes of the organism. My emphasis is simply this: I began by saying how important the invasion criterion was. All organisms that are proper species have the ability to increase when rare. It seems to me that if you are interested in risk, you ask where is it going to increase and that has to do with environment, not with species. So, from an ecological point of view the only place to start is to look at the interaction between the organism and the environment. If everybody has given the emphasis to the organism, I have given the emphasis to the environment. But I am not trying to polarize it completely.

Intervention:
So, it is in balance; we should look at the characteristics of the organism, at the new traits and at the environment and at their interactions.

Intervention:
Can I just say that if you look at the case history you used in the report, *Spartana anglica*, that there several levels of explanations of the success of that species. You might be interested in the fact that it was an allopolyploid or that it might have certain fixed levels of heterozygotity. Or you might be interested in its unique physiology. Or you might be interested in probably the real reason it did so well, namely, that it can grow below any other species in temperate zone salt marshes. There is a hole in the environment; the niche is there. I think the difficulty is that you can see the environment that might be invaded but you couldn't necessarily see that that plant would produce the necessary characteristics to invade it. I might also add, since we seem to limit long term to one century, that that species is still evolving, that it is genetically uniform, that it is being attacked by a certain fungus, that it is retreating in parts of its range and that it is clearly taking on a new ecology. We are still learning from a species which was actually born one hundred years ago in 1892.

Intervention:
My question is how can we deal with unpredictability? We have heard that predictability depends on the organism and on the environment and we have a lot of environments outside. So, how can we handle this? Most of the real risk assessment can't involve all environmental effects. Therefore we should look at monitoring and its design. In this field, I think, we should focus on the GMO, on target-organisms but also on the environments where releases are actually taking place. It would be very helpful to the next generations of ecologists over one hundred years, when we have the long term effects, if they have at their disposal the data we now select from the releases. In my view, the more data we take now from the release sites, the better it would be for the next generations of ecologists.

Intervention:
In part as a response to that, we are making an assessment at the moment about releasing something into the environment. We use monitoring data extending over hundreds of years in relation to all the crop plant species which are being modified. We know historically from agricultural work that all the various crosses that have been used have thrown up enormous numbers of variance. And we have experience from breeding programmes and from farmers' use of those crops as to what is happening. So, there is basis of experience based on monitoring not done for its own sake. You might make the comment: Why don't we do more monitoring with the existing experiments? But the point has been well made that they have been so small that they are not likely to yield anything of consequence. And in part this is because the number of propagules that have been released into the environment is so small and they are not being released repeatedly year on year over very large numbers of environments. The amount the work involved to do very large size experiments over a large number of areas is not feasible financially or in terms of the number of people available to do the type of monitoring that might be required. And also, as has been said before: What do you monitor? What do you look for? If you see a small change in one thing, to what do you relate it to? Was it a particularly wet summer that year?

Question:
Herbicide resistant plant are not a topic in this conference. Is this because ecologists agree that engineering herbicide resistance is not ecological sound? Why is this topic not discussed?

Response:
We had about thirteen to fourteen new traits on our list we would like to have be discussed. We had to limit ourselves to five topics. We simply had to select. Nevertheless, herbicide resistance is one of those issues always coming up in these kinds of discussions, I would be happy if those of you who call themselves ecologists to respond to this.

Response:
I find it very difficult to see why -pretending that I was an ecologist- I should be worried be by the escape of herbicide resistance genes into indigenous flora. I can see that there are second or third order problems with introducing herbicide resistance genes into crops which has nothing to do with the particular technology that was used to generate the herbicide resistance. It has far more to do with patterns of herbicide use and whether that is appropriate in agriculture. But the escape of herbicide tolerance genes into indigenous plants I personally don't see as any danger to the natural environment. Although I can see it being of an embarrassment to both the producers of the herbicide resistant varieties and somewhat more costly for some the poor farmers who are going to be at the receiving end. But it is not going to generate a problem to which there is not already a solution. It is not a novel solution: Change herbicide. I might add that herbicide resistance does occur naturally and has been causing problems in some areas anyway. Basically, I don't think that herbicide resistance is a problem. I actually will look forward to the

commercialization of herbicide resistant crop varieties in Europe because I would love to see the scale of the experiment that would be carried out there and to see the rate at which these lovely marker genes do actually get out into the environment; we will have a very large body of people who are monitoring this experiment and they are called the farmers; they are not going to miss anything.

Intervention:
The point I wanted to make is that we seems to step back from the discussion the minute it looks it is going to be a problem for agriculture. The fact that herbicide resistance genes, virus resistance genes and insect resistance genes sitting in wild plants are not being of selective value means that they presumably can come back into the crops at a later stage. Crops have an ecology; ecology is not about the natural environment entirely. I don't think that ecologists should stop when saying: Well, it is not a problem for the natural environment. The environment includes that which we farm.

Intervention:
Recently, we had a discussion in The Netherlands on the use of herbicide resistance genes in crops with all groups involved in this problem; not only ecologists but also consumers and industry representatives. It was quite a surprise how quickly and how fully we reached an agreement. The real point is the usage of herbicide and not the herbicide resistance genes. To my opinion, that is very rational point of view.

Question:
I don't think that my question about herbicide resistance has been answered. But I don't think it can be answered that easily. In Germany, for example, we had a huge conference on herbicide resistant plants. The results somewhat differed from those that were mentioned here. Can I raise another question? I would like to know how you look at the situation when very many genetically engineered crops are being commercially grown and genes are being transferred to wild flora in rather big numbers. I can imagine that this will impact processes of speciation in wild plants. Could somebody elaborate on this point?

Question:
It seems that herbicide resistance genes have received a kind of approval. Are we now moving into a direction that genes are approved and therefore can be used in larger field trials or in production scheme? What about synergistic effects?

Intervention:
I am a little bothered that you put all herbicide resistance genes together because there many different types of herbicides and many different modes of action. And when you are saying herbicide genes get a stamp of approval, I think you need to be cautious on giving a blanket endorsement all of them. That is dangerous. I myself sort of fall into the camp that I would be glad when some of herbicide resistant plants are used, too. I feel that the role of herbicide resistance genes in wild populations is not that much of an issue because these are areas where herbicide

treatment is not used. And I don't think that the selection pressure will be there; those genes in the wild plants therefore will probably disappear.

Question:
Could you take one step further because we are talking not only about herbicide resistance genes but also about many more types of genes, like selectable markers, insect resistance genes and so forth? Is it inevitable in the future that due to breeding programmes many of those kind of genes could come together.

Response:
Again, I think when you are looking at natural populations of plants, you are looking at a very diverse environment, whereas these plants, into which these genes are being introduced, are essentially grown in a very highly managed environment where you have series of the same plant. Again, I think it is going to be selection pressure on whether or not those genes or those sets of genes will give plants a selective advantage and whether they are going to be maintained in the population of wild plants. The original entry is going to be different than maintenance over a long term. I don't know if you could design a series of genes that if you wanted it to move into a particularly tree species, whether it would make that species more invasive -I suppose you could speculate about that. In addition, I also think the probability of that happening would be extremely low, given the specificity of transfer by genetic means to wild plants which is really the only way that we know it could happen now. In the crop plants where these genes are being introduced, with few exceptions, there are not wild relatives around there.

Question:
I think professor Schell addressed this issue in his introductory comments when he mentioned that human beings have been mixing many unknown genes for probably three thousands of years or more. I mean we have had many long term experiments going on. Is there really anything going to be different here, especially if we know many detailed properties of the individual genes involved?

Response:
Basically, I say on all occasions where we have wild relatives of cultivated crops, then ever since those cultivated crops have been grown alongside those wild relatives. If there is any possibility of sexual compatible gene transfer, it will have been occurring. There is certainly fairly good evidence of it in the instances where it has actually been started. For example, there was an area in Central America that was particularly rich in germplasm for *Zea mays* and it was essentially a series of landraces, loosely speaking in the region of Mexico, and this was being kept as a nice genetic resource which maize/corn breeders could go back to if they wanted. When they did go back and looked at these landraces, what they found was that the genetic diversity that had existed a decade earlier had now been dramatically reduced and then replaced by genes which are characteristics for modern varieties that the Mexicans were growing in that area. So, there clearly had been a lot of gene transfer from a crop into those landraces. So, this was very sad; you had an erosion of a genetic resource. The point I wanted to make is that crop genes have been

introgressing into wild relatives for about the last seven thousands of years in the case of wheat. And yes, it will almost certainly have had an effect on the ecology of those species that were receiving those crop genes. But basically the species that we are looking at now, which are relatives of crop plants, are the way they are now, at least in part, the result of the genes that they have been receiving from those cultivated crops. It may be good or it may be bad; we don't have the control to tell. We just have that situation. If we then go to ask what I think was one of the aims of the question you were asking, is there going to be any undesirable effect of the transfer of genes of genetically modified crops into wild relatives?; Is it going to affect speciation or whatever?, then I would need to ask the question: Do we think there is something qualitatively different about the genes which have been inserted into those genetically modified crops which would predispose us to think that it would have an effect on speciation? I am afraid that I, from my position of informed speculation, can't see why, for example, herbicide resistance genes are likely to affect speciation.

Intervention:
Yes, it is very nice to think that we should preserve as much as genetic variability as possible. But if we look back over the last few centuries in Europe and South East Asia and may other parts of the world, we see that have altered the environment so enormously that we have eliminated an enormous amount of genetic variability and the ability of existing species to change. We cannot alter that; there are going to be more people and we don't have the luxury to allow the native species around us to live as they might have lived in God's perfect evolving world. We have adjusted that and we are in control of it whether we like or not! It is quite clearly important that we don't make stupid, foreseeable mistakes but we cannot step back; we have done the damage.

Intervention:
I would like to comment on the remark that we are using well known genes which we are now introducing into plants. In some cases those genes are from bacteria. A couple of days ago I have been at a meeting on bacterial genetics and ecology and there it became very apparent that for most genes of bacteria we do know what their functions in the laboratory are but we absolutely don't have a clue what their influence is on the ecology of the bacteria. The plasticity of bacterial populations is really enormous and nobody has any idea what genes have what influence. This is not to say that the situation is that dramatic when you put bacterial gene into a plant. For example, you are able to predict what a kanamycin resistance gene is going to do. But in general I don't think we should overestimate what we know.

Intervention:
I would like to add it is not only a question from which genepool a gene stems but also at which position it is integrated into the genome and for a bacterial gene there is not a right site in the genome of a plant. There is another point I would like to make. I am not talking about stepping back but what we are talking about here is stepping in front. And I think, when we as human beings want to step in front, we have to argue and to think about if we want to do this or not. It is not a natural

evolution but a man-made evolution. So, I would like to make a differentiation between stepping back and stepping forward. I also would like to add a difference between resistance genetic engineered into a plant and resistance mechanisms observed in wild populations; the first one is mostly based on one gene, whereas the latter one rely on multiple genes and this may cause differences in overcoming those resistances.

Intervention:
If you take the example of a virus infecting fields of peaches, apricots and Japanese plums taking place all over Europe, and you hear that all the apricots need to be eradicated because they are the most sensitive crop to the virus while the other species can withstand, then you ask yourself what other alternative do you have at the moment to conserve apricots. Or do you want to cut all the apricot trees and that is the end of the story? There are no resistance genes in the gene pool of apricot available as you would like to have them. So, the first step will be to go to the viral genome and look there for sequences which you might integrate. Is there an alternative to this or do you really say: I don't care about apricots?

Intervention:
I want to talk about what we can know about genes. I think that is an important point. And I think that a gene that we introduce into the plant is generally made up of several pieces of genes from different organisms, so that when you take a segment of the bacterial sequence it is generally together with viral or plant sequences in order to allow its expression. Actually we do know some of these genes extraordinarily well. Maybe they are among the very best known genes that exist and very often much better than natural genes. We have been able to determine what the gene product is, that is to say what the protein and its function is. We have been able to study where in the plant the gene is expressed and how and under what type of control. I think that type of knowledge is something that we really should not dismiss. In a sense this type of knowledge has enabled us to be here and discuss those sort of questions.

I was very glad that you brought up the issue of resistance genes because I felt there is quite a lot more to be said about that. I think that everyone who is working with resistance genes is abundantly aware of the fact that resistance genes have a limited life-span in plant breeding in general and we are convinced that the only durable types of resistance that we will be able to develop over the medium range will have to be based on multiple genes strategies. I think that this was obvious in the discussions on insect resistance. In addition, I think one should not be worried about the extinction of the classical breeder and classical virologist because what we are doing is just contributing a few more genes that can be used in a classical breeding process, adding together partial resistances.

Intervention:
I just would like to reiterate what has been said about the apricot. It seems to me the key question is a more comparative approach. I don't have regulatory experience but I worry a little bit about the case-by-case approach if it is used to exclude a

comparative approach. It reminds me of the story of two travellers who disturbed a grizzly bear. One of them proceeds to buckle on his running shoes and the other one says: You are wasting your time because you cannot outrun a grizzly bear. He says: I don't have to, I just have to run faster than you. This is the point: We won't have all the answers on our risk assessment but we can reasonably say the new answer that has been proposed should have obvious merits over the existing answer, when you are faced with many existing practices which are extremely unsatisfactory. I am very keen on the herbicide resistance cotton because in the United States cotton cultivation constitutes about 40 % of the current pesticide burden. I am very interested to see whether the people in Kazachstan and Uzbekistan will switch form the currently appallingly destructive use of pesticides for their cotton cultivation there. The example I found most striking was at a workshop on vaccines a couple of weeks ago when two researchers were describing the point mutations on which existing poliovirus depends; you would never authorize such a vaccine if you had known it depended on a single point mutation. The genetic engineer can offer you a poliovirus vaccine from which he deletes a mass of amount of a sort of deletions that you saw in the NIVAC example of the vaccinia vaccine. We don't know all of the ecological answers about its impact but it is patently more safe than the existing practice. So, I don't think we should dilute ourselves about absolute ability to predict. I think we should put much more emphasis on a comparative aspect. We are in an unstable and threatening situation, so that in some sense I want to interpret the step-by-step approach almost as meaning: We should back off this massive mixing of unknown gene pools and only allow agricultural innovations that put in one characterized gene at the time. Therefore, only GMOs should be introduced in agricultural innovation.

Question:
This morning I learned from dr. Kimman's presentation that using a recombinant vaccine might save an African dog from extinction. Could we discuss equivalent cases in plants like we just heard about the apricot? With regard to genetic resources I would like come to back on the issue of genetic diversity; I don't know whether this is the right forum but I would like to hear something more about it.

Intervention:
I think that the gene that one would need to control the pressure on plants and plant genetic resources is something which controlled man.

Intervention:
Are we going to save wild species by genetic engineering?...I would answer that by saying: We have greatest difficulty in maintaining agricultural productivity in areas that it matters.

Intervention:
I would like to go back to that very good point that was made about gene expression and the siting of genes. I have a conceptual problem at the moment that there are people talking about gene position and gene expression. My own feeling is that our understanding of how to express genes properly in target tissues in appropriate

amounts is so poor that if we are really going to approve the release of anything, we must only approve the release of genes that can be expressed at extremely highly levels and still be safe. So, as a result their is no need at all to worry about where they are located. Having said that, I do see a long term serious problem that we are surely going to move towards plants where there is gene expression of toxins in target tissues which are not eaten to control insect pest. When are we going to have sufficient knowledge of gene expression and regulation that we can safely do that. And for those of you who think you can do it already my question to you would be: Would you willingly eat your organism if it was ricin that coded for in one of these highly specific promoters? Because until you have that feeling of safety, I don't know whether it can be safe to let these things out. And yet we have very toxic crops; potatoes theoretically could kill us if the natural gene expression wasn't so well regulated.

Intervention:
I just wanted to say that scientific research is a step-by-step process as well and as regards to control of gene expression I wouldn't eat that plant expressing the ricin gene. All of us working on gene expression know that there are more surprises everyday and it is absolutely true that we do not master that aspect of transgenic plants. I think that the whole field that has been very loosely called 'position effects' which have to do with some unexpected regulatory phenomena in general is one of the fields which is exploding. You can see exciting papers completely blowing this whole field open almost on a weekly basis. This also reminds me of an example of 'toxic' potatoes which were made by conventional crossing and were certified and cultivated and then it turns out that under certain climatic conditions the alkaloids that normally were only produced in the leaves were also being produced in the tubers. This is an excellent example where crossing two genotypes of potato led to changes in gene expression which led to an unexpected change in alkaloid production. So, this is a classic example of the unpredictability of mixing enormous numbers of genes, even in species that we think we know rather well.

Intervention:
Maybe we should only approve those new varieties of which we know the molecular biology. A lot of plant breeding throughout history has been to detoxify the wild plants. There are a lot of plants that we eat of which their wild relatives are toxic. And the breeders have developed these plants, so that the varieties not only produced better but are also consistent and are non-toxic as food or feed. When the breeders make crosses for insect or disease resistance, they don't have the foggiest idea why they are resistant. But we still think that is the way to do it and we also know very little about what the mechanisms of these genes are. In response to the question why don't we use multiple gene based resistances, we don't even know how to handle single genes. I think as we learn more about single gene resistance or at least something that appears by Mendelian inheritance to be a single gene, then maybe we can learn to go on with multiple genes. So, we still have a great deal to learn.

Intervention:
I just would like to say that using of single genes might be dangerous. As regards to the Bt cotton, the farmer has to use an insecticide as well; on the short term a farmer who thinks he will have a better yield is not going to spend more money on an insecticide if he is going to have the same yield anyway for the next three or four years. I feel that this Bt cotton that has been called a safe transgenic plant is only safe if certain guidelines are followed. I don't think you can call it a safe transgenic plant.

Intervention:
I assume the safety you are talking about is safe in terms of controlling insects. As far as I understand the reason of the use of insecticides in combination with the Bt is to get essentially a quicker kill because like baculoviruses Bt, when used in a bacterial form, is rather slow in killing. So, the use of the biologicals will reduce overall level of populations of insects but won't kill immediately like chemical insecticides. So that when the infestation is high you can come in with one insecticide application and then let the Bt to do the rest of it. Now when Bt is incorporated into the plant, the action is a little bit more rapid since the insects essentially quit feeding, so that the Bt in a plant does not require the additional application of the insecticide.

Intervention:
What about virus resistant plant? We heard that viruses mutate rapidly and may overcome resistance quite quickly, so therefore it would surely be better to have a sort of double-cassette in the plant. Is it worthwhile to spend so much money on a plant when its resistance can be quickly overcome?

Intervention:
Here we face the danger of entering another discussion because you are referring to that your product is not as valuable as you though it was, whereas here we are talking about unexpected or negative ecological effects.

Intervention:
I think that we should not give the impression that we know that the virus resistance genes that we use are going to be overcome. I think that we are all talking about that possibility because we need to start to think about in case they are overcome. The people in the field discuss the question endlessly. I would like to mention the example of certain coat protein mediated resistances surprisingly also providing protection against a wide range of viruses. This is used as an argument: Well, if we have viruses which only have 60 to 70 % similarity, this means there is a fair amount of mutational looseness that can be accepted in overcoming a challenge infection. So, a lot of people consider that this means there is good chance that some of the virus genes may in fact be very durable. But to be honest, only in ten or hundred years we will know what the durability of these various types is. That is also why labs around the world are developing as many different strategies of virus resistance as possible, so that if and when it becomes necessary they can be

combined and changed. But I don't think that we can conclude that they will be overcome.

Intervention:
I have a philosophical problem. I think we have discussed the kind of short term problems we regularly discuss in our biosafety committees. And I think that they are not really commensurate with the long term problems which are the reason why we sit here. What I am looking for is a kind of measuring stick. How important are our short term problems in the long run? To find a measuring stick may I propose to look at the resilience of ecosystems and to see whether we can define resilience of ecosystems in the long run. I try to give you an example. We had an agro-ecosystem in Europe where tobacco was grown -it is an important cash crops in several countries and governments are making money out of that tobacco, then a terrible disease appeared; the tobacco blue mould and tobacco was nearly wiped out. Within ten years, however, the tobacco industry was on its feet again. Here we have a terrific resilience of a man-supported agro-ecosystem. You can also talk about the major perturbation in Europe; the introduction of potato. Another major perturbation in Africa was the introduction of maize. Once the maize is there it is incredibly resilient as an ecosystem; new diseases were introduced and they were overcome by multiple resistance, just due to selection by farmers, no plant breeder participated in that process. Apparently it is resilient. But Africa itself is also resilient because when you stop maize growing today, then it disappears and they original forests will resume. The same is probably true for Europe: If we stop potato growing, then a forest will appear on the fields. So, the question is: Can we define resilience in the long run? And then use that to measure the importance of our short term concerns. Then we have really contributed something to the issue of the conference. But this is a philosophical question: I don't know the answers.

Intervention:
We talked about long term effects but we did not define what we see as unwanted long term effects. We are human beings and we alter our environment. Perhaps we should define what kind of environment we want to have over one hundred years and then compare this to what we are doing now.

Intervention:
A great deal of research work has been done on resilience at the ASA, the international institute for applied system analysis. They are involved in a study using a three hundred years time based backward and a three hundred years forward approach for projection with links to the international biosphere and geosphere programme. So, there is work in this field on the definition of long term resilience.

Intervention:
It is not only what we want to have in hundred years but also what we need to have. We should take care that we want what we need.

Intervention:
How do we get a consensus of deciding what we want in hundred years? Do we have an European Community Directive?

Intervention:
Ecosystem level resilience is one of those things which is extremely difficult to quantify. The problem is for instance that you can have communities that change radically but which have exactly the same measures of resilience but those ecosystem level standards. So, on the one hand you have an ecosystem that you want because it is useful to you. On the other hand, you have an ecosystem which is completely useless, even though the indicators are exactly the same. So, quantifying and maintaining ecosystem resilience could be seriously bad news.

Intervention:
I thought that I learned from many risk expert who are saying that in our time we cannot use a complex technological system any more just because of the risk qualities. And if I compare the discussion to this, I see that we have no solution for the long term handling of genetically engineered systems.

Especially the ecologists tell us that this system is not predictable, it is a too many multiparameter system, it is too complex. So far we have looked at the actual risk and I think we had some good guidelines in the EC but, and that is my feeling about this conference, with the long term aspect we have a completely new situation in the calculation of possible risks. And if I hear all the experts present here, I am not sure any more whether we can really handle those genetically engineered systems.

Intervention:
You are absolutely correct. We are talking here about uncertainties and unpredictabilities. However, many of us say that you have to ask yourself what the result would be if you don't do it. We had a discussion in the parliament of The Netherlands. Somebody raised his hand and said that when there is any, even very remote chance of a risk, why don't we simply forbid it? Simply because there may be a lot of beneficial application. The problem is to pick out those applications where we can foresee that something might go wrong.

Intervention:
Whenever we are inventing a new hybrid plant made by very classical breeding methods, there are no risk assessments made.

Intervention:
We never really got to discuss the antifreeze gene in the range extension of a top predator. And if you want to force an ecologist like me to say something unequivocal, then I should say: Ban it! Don't even think about it because that has very serious large scale implications on the long term. All the other things we have talked about to an ecologist are likely to be very slight, if any, long term environmental impact, by which I mean herbicide resistant plants, insect resistant

weeds and so on; all these are of a completely other order than extending the range of a keystone predator.

Intervention:
I would like the following issue to be addressed; the question of long term large scale use of different crops with multiple traits inserted at the same time as biological pesticides with different genes inserted and a whole lot of other things. In many cases there may not be particular synergistic effect. And at the moment we are looking at specific crops with single genes in an isolated situation. But one should look ahead.

CONCLUDING REMARKS BY THE GENERAL RAPPORTEUR

Dr. R. Harrington, Conservation Genetics and Mammal Research, National Parks and Wildlife Service, Ireland

Targeting the conclusions of this conference is indeed a challenge given the diversity of interests and disciplines of the participants and the wide range of topics that have been presented. Nevertheless, the conference has been singularly cohesive by recognizing and working within the continuum that exists between molecular biology to that of ecology. The approach of the molecular scientist conferring or working with the ecologist emphasizes the unity that exists in biology and specifically the fundamental relevance of evolution; to quote Dobzhansky (1973) one of the foremost geneticists of this century, "Nothing in biology makes sense except in the light of evolution." This statement would seem to be central to our discussion as we ponder the potential long term ecological impact of GMOs. It implies the need to establish an understanding of evolutionary pathways and patterns that belie the interaction of all organisms and their environments. Specifically relevant to this conference, is whether the GMOs acquired trait/s can alter their competitiveness, resistance or tolerances that establish the parameter values of GMO populations in a range of habitats.

The significance of the distinction that has been made between recombinant genetic exchange and that by "natural" processes, e.g. sexual methods, is as yet, not clear. Recombinant techniques allow the transfer of genetic material between distantly related organisms. Given the universal structure of DNA, genetic exchange can now occur between prokaryotes, eukaryotes and viruses alike. This is not the case with "natural" processes where genetic exchange is limited to between relatively closely related organisms. Another distinction that has been made between recombinant and "natural" processes is that of the considerable differences in the amounts of genetic material that is exchanged within each of these two processes. In addition, it has been pointed out that the level of knowledge of the genetic structure is of a high level in the case of the GMOs traits and of a low level in the case of "natural" processes. Whether these differences have any significance with regard to the parameter values of any organism would appear to depend upon the acquired traits/s and the environment into which it might be released.

The concept of keystone species mentioned by dr. Crawley introduces the possibility of identifying and categorizing organisms, which if altered by recombinant techniques, that might give them the possibility to extend into habitats from which they had previously been excluded, resulting in significant adverse impact on those habitats. However, the question as to whether a single or a small number of recombinant genes could significantly alter the genetic structure of such organisms to enable them to invade new habitats requires further study. It has been suggested for example that a keystone species such as the Atlantic salmon (*Salmo salar*) could have a major ecological impact if given cold tolerance genes that might allow it to invade the Arctic waters; but is this the only limiting factor preventing salmon from entering and establishing populations in the Arctic?

The subject of invasiveness of organisms has been closely linked with that of predicting the long term ecological impacts of GMOs. This linkage was developed by steering committee organizing this conference, by the Council of Europe document (Potential long-term ecological impacts of genetically modified organisms, 1993) and during this meeting. However, predicting how organisms might behave in a particular habitat is extremely difficult. Nevertheless, ecological insight provides the only basis to explore the potential impact of exotic organisms and probably GMOs that might persist in the unconfined environment. Such insight can never be absolute or precise given the great complexity of ecosystems. Should ecology be integrated into the role of risk assessment it could help foresee some of the adverse impacts of GMO technology. Indeed the role of ecology could further extend the application of the use of GMOs to nature conservation itself as demonstrated by the work of professor Pastoret with vaccinia-rabies virus.

The role of monitoring as an aide to predicting the impact of GMOs on the environment is restricted by being only extrapolative. In addition the role of monitoring GMOs has been shown by dr. Gliddon to be as yet inoperative as most if not all releases, that are known, are confined to the release site and are of short duration. He also explained how difficult it is to know what to monitor for. Using the example of introgression between *Senecio vulgaris* and *S. squalidus* he described how unpredictable gene flow can be; not only between what was considered as distinct species but also between subsequent populations that had acquired their own genetically distinct status. Nevertheless, he recommended three tasks to enhance the effectiveness of monitoring:

1. Identification of GMOs that have the potential to impact on the environment;
2. Consideration of the fitness of the GMOs that are released, and;
3. Rates of spread of genes through wild and feral populations of related species.

The first two of the above mentioned recommendations are difficult without insight into the evolutionary background of the GMOs and the ecosystem they are candidates for release in. The third recommendation requires the application of considerably higher levels of attention to population genetics than what generally applies at present. The capacity to monitor is further limited by resources. Considerable merit was shown in utilizing the non-scientist especially the rural dweller and farming community generally as additional detectors of changes in the countryside to which GMOs had been released. It would seem that there is even greater merit in incorporating aspects of GMO monitoring into various ecological change monitorings that are already conducted in a number of countries, e.g. The Environmental Change Network that is co-ordinated by the Natural Environment Research Council in the United Kingdom.

The lack of any fundamental ecological distinction between natural and man-modified environment was made repeatedly during the conference. Thus emphasizing the need to accept the linkage that occurs between human activity and nature in all its aspects. Once GMOs are established in any un-closed environment they are unlikely to be eliminated, the ensuing gene flow only recognizing the reproductive compatibilities that apply, whether in the wilderness or in man-modified environments, e.g. farmland.

Socio-economic considerations arose a number of times and we stopped short of discussing them in any depth. However, the most significant long term impact of GMO release may have more to do with the scale of their use, through the use of crops in areas from which they are now excluded, further intensifying crop production or further increasing dependence on technological development. However, GMOs can not be un-invented. Indeed they may be the most environmentally acceptable way of addressing many of the world's human population needs. Whatever the course for the future the most important consideration for us, our institutions and governments is to recognize the role that biology has to play in helping to make the best use of GMOs; recognizing both, the short-term, often economically and commercially oriented aspects of molecular biology and that of ecological and ecosystem study that are generally long-term and less obvious for funding and support.

The philosopher, scientist and administrator Goethe, who lived in Strasbourg, said that there is no idea more powerful than one that has come of age; and so it would seems, as our human society endeavours to make the best use of GMOs, recognizing the potential of the biological continuum that exists between molecular biology and ecology.

REFERENCE

Dobzhansky, T. 1973. Nothing in biology makes sense except in the light of evolution. Amer. Biol. Teacher (March): 125-129.

[illegible] GMO strains established in an open environment, they are unlikely to be eliminated [illegible] [illegible] [illegible], e.g. Scotland.

[illegible] [illegible] [illegible] [illegible] [illegible] [illegible] [illegible] [illegible] [illegible] However, [illegible] [illegible] [illegible] [illegible] [illegible] [illegible] [illegible] [illegible]

[illegible] [illegible] [illegible] [illegible] GMOs [illegible] [illegible]

REFERENCE

[illegible] (March) 1991 [illegible]

POSTER SESSION

ABSTRACTS

FIELD TRIALS OF TRANSGENIC PLANTS WORLDWIDE

P. Ahl Goy and J. Duesing, Ciba, Seeds Division, Switzerland

The number of field trials with genetically modified plants worldwide has risen concomitantly with the growing use of biotechnology for crop improvement. Up to the end of 1992, 675 field trials have been conducted, where one trial is defined as "a single basic strategy introduced into a single crop and tested in a single country (one or several locations), by one company or institution in one year". The total number of trials will be around 1000 on over 2000 sites by the end of 1993. Most of the activity to date has occurred in North America (48%) and in the EC/EFTA (42%), with the rest being split between Central and South America (7%), the Pacific Rim (3%) and the Middle East (<1%). The private sector accounts worldwide for about 80% of the trials; only in the Pacific Rim has the public sector been responsible;e for a majority of the trials (84%). The yearly number of trials is still increasing in North America, while European activity appears to have stagnated since 1991.

Today, 33 different species have been tested worldwide: potato, oilseed rape, tobacco, tomato, maize, sugar beet, cotton, soybean and alfalfa, in decreasing order, represent 90% of all trials. The most commonly evaluated trait is herbicide tolerance (36% of the trials), followed by improvement of product quality or production methods (17%), virus resistance (15%), insect resistance (13%) and marker genes (12%). Only a few trials involved fungal or bacterial resistance (3%). With many companies entering a development phase towards commercialization, there is now a clear shift from model plant species and marker genes to agronomically important crops and traits. Further changes in the trends are expected: new genetic strategies are developed and other crops (e.g. wheat and rice) are routinely transformed.

RUNNING WILD OF TRANSGENIC SUGAR BEET: COMPETITIVENESS AND HYBRIDIZATION WITH RELATIVES

Dr. D. Bartsch, Department of Biology, Technical University of Aachen, Germany

The competitiveness of vegetative sugar beet genotypes (*Beta vulgaris* var. *altissima*) is demonstrated in auto-ecological field experiments on base of biomass production under competitive stress with fat hen (*Chenopodium album*). Besides containing a gene construct leading to BNYV-virus resistance, the transgenic sugar beets also contain genetical markers for herbicide and antibiotic resistance. Field experiments are done in Germany at a site with and a site without BNYV-virus infection. The experimental program focuses on the investigation of ecological interactions of sugar beet in the target ecosystem. Young fallow is the most interesting area in the target-ecosystem where an unwanted survival of sugar beets could realistically occur. Sugar beet is a relatively weak competitor and, at the same time, is dependent on the high nutrient levels present in the test field. The growth of the sugar beet will be assessed in a situation where it is competing with a therophyte typical for root crop culture. Parallel, the sugar beets are hybridized with its relatives beetroot (var. *conditivia*), spinach beet (var. *vulgaris*), and wild beet (subsp. *maritima*) to simulate the consequences of gene transfer. The competitiveness of the new hybrids will be tested in greenhouse experiments in 1994. The project is intended to improve our understanding of aspects of biological security and ecological risk assessment used for the environmental release of genetically modified plants.

COAT PROTEIN MEDIATED PROTECTION AGAINST PLUM POX VIRUS IN HERBACEOUS MODEL PLANTS AND TRANSFORMATION OF APRICOT AND PLUM

A. da Câmara Machado, H. Katinger, M. Laimer da Câmara Machado, Institute of Applied Microbiology, University of Agriculture and Forestry, Austria

Improvement of fruit tress through traditional breeding methods is a long term effort because of their long generation time. Thus, new approaches are needed to reach the envisaged breeding goals in a reasonable time frame. Genetic transformation potentially is useful, because specific changes can be made.

In the last few years the expression of the viral coat protein gene in transgenic plants has been shown to induce tolerance against virus infection. Transgenic plants of *Nicotinia clevelandii* and *Nicotinia benthamaiana* -herbaceous host plants for PPV-transformed with *Agrobacterium* strain LBA 4404 containing the plasmid pBin PPVm, regenerated on selection media containing kanamycin were tested for the expression of the PPV-coat protein gene by ELISA and immuno western blot. After rooting and acclimatization plants were tested for the protection against PPV. Following the inoculation plants were investigated for symptom development and virus accumulation. Different lines were identified, according to the different reaction to the mechanical inoculation ranging from a complete absence of symptoms to a strong reduction.

There have not been many reports on trees in general and fruit trees particularly. It is obvious that the major obstacle for the transformation of fruit tree species is the regeneration of transformed plantlets. Attempts to improve crop plants by genetic engineering techniques will always depend very strongly on the availability of reliable protocols for transformation, selection and regeneration. Different systems have been developed allowing the transfer of foreign genes into apricot and plum cultivars. We report the transformation and regeneration of *Prunus armenica* and *domestica* plants with *Agrobacterium tumefaciens* strain LBA 4404 containing binary plasmids, pBinGUSint, carrying the marker gene β-glucuronidase (GUS) and pBinPPVm, carrying the coat protein gene of Plum Pox Virus (PPV). The marker gene GUS was used for the optical evaluation of the efficiency of the transformation system. The coat protein gene of PPV was used to introduce the coat protein mediated resistance against one the most important pathogens of stone fruit trees in Europe and the whole Mediterranean area.

GENE TRANSFER AMONG RUMEN BACTERIA

K.P. Scott and H.J. Flint, Rowett Research Institute, United Kingdom

The rumen harbours a very large and diverse microbial community comprising anaerobic bacteria, protozoa and fungi. Ingested micro-organisms are able to intermix directly with the endogenous rumen microbes since there is no preceding acidic barrier. These factors make the rumen a potentially important site for the horizontal transfer of genes, including antibiotic resistance determinants and recombinant DNA sequences, between diverse micro-organisms. Genetically manipulated micro-organism may in future be deliberately introduced into the rumen as probiotics, or may enter the rumen as a result of their use in silage treatment or other environmental application.

We have found evidence antibiotic resistance genes among the endogenous rumen micro-organisms, but the transfer mechanisms by which they were acquired are not yet known. We are looking specifically at types of tetracycline resistance genes found among rumen anaerobes, and the genetic variability and spread. In addition, we are investigating the occurrence of gene transfer between rumen micro-organisms under rumen conditions.

ECOLOGICAL MODELS OF CONVENTIONAL AND TRANSGENIC CULTIGENS BECOMING FERAL

U. Sukopp and H. Sukopp, Institute of Ecology, Technical University Berlin, Germany

Introduction

If transgenic cultigens are released into the environment they may become feral as we know of many conventional cultigens. This process can basically happen in two ways: through hybridization with closely related wild plants (Table 1) or through revert to the wild-type.

Table 1: Cultigens becoming feral with closely related Central European wild plants as potential hybridization partners.

CULTIGENS	WILD RELATIVES
Beta vulgaris subsp. rapacea Beta vulgaris subsp. vulgaris	Beta vulgaris subsp. maritima
Brassica napus Brassica rapa Brassica oleracea Brassica juncea	Brassica oleracea Brassica nigra Raphanus raphanistrum Sinapis arvensis Hirschfeldia incana
Setaria italica subsp. italica	Setaria italica subsp. viridis
Daucus carota subsp. sativus	Daucus carota subsp. carota
Medicago sativa	Medicago falcata
Malus domestica	Malus sylvestris
Pyrus communis	Pyrus pyraster
Populus x canadensis	Populus nigra Populus alba
Lolium perenne (cultivated forms) Lolium multiflorum (cultivated forms)	Lolium perenne Festuca pratensis
Trifolium repens (cultivated forms)	Trifolium repens

In order to predict the probability of transgenic cultigens becoming feral we can make a direct approach by investigating the transgenic cultigens themselves in the field with a great effort of time and money. Unfortunately these experiments - limited in time and space - are not suitable to predict the long-term ecological behaviour of transgenic cultigens if they are planted at a large scale.

Due to a virtually complete lack of information on the long-term ecological behaviour of transgenic cultigens the authors suggest to make an indirect approach. This requires adequate ecological models that are based on a wide range of statistical information. Our poster presents two different ecological models (Figure 1) to predict the ecological risks of transgenic cultigens.

Model 1

If we make a prognosis about a single, precisely characterized, transgenic cultigen it makes sense to compare it directly to the original plant that was conventionally bred and that we are familiar with (Figure 1, Model 1). In a so-called "case by case study" a conventional cultigen functions as a model for a transgenic cultigen and provides us with information on the probability of a certain transgenic cultigen becoming feral and on possible undesired ecological effects of this process particularly in man-made ecosystems. If we see a great risk of a certain conventional cultigen becoming feral - e.g. through hybridization with closely related wild plants (Table 1) - we should assume the same risk of the derived transgenic cultigen.

Model 2

However, if we try to predict undesired ecological long-term effects in semi-natural and natural ecosystems we suggest to compare the process of becoming feral in general to the introduction and naturalization of non-native species (Figure 1, Model 2). The intended or unintended introduction of non-native species by man and the subsequent naturalization of some of these species (Table 2) functions as an appropriate and valuable model of the long term ecological behaviour of both transgenic and conventional cultigens that have escaped cultivation and invaded semi-natural and natural ecosystems. Model 2 has been described by Sharples (1982) and Regal (1986, 1993) as the "Exotic Species Model".

Table 2: Native and non-native flowering plant species in some European countries.

	Total number of flowering plant species	Percentage of natives	Percentage of non-natives
Germany	2147	80%	20%
Finland	1227	82%	18%
Poland	2250	87%	13%

Model 2 does not state anything about single cases, but sets the pool of introduced and naturalized non-native plants equal to the pool of cultigens without making a difference between conventional and transgenic cultigens. Because of the biological mechanisms of plant dispersal we must consider very long periods of time. Therefore a data base ranging far back in history is used in Model 2.

Both transgenic cultigens and non-native plants are released in a limited number of eco-systems (Figure 2). Subsequently they can spontaneously spread to many other ecosystems -in future even to currently not existing ones. This makes clear that a short-term perspective is not adequate and that we should use the long-term data base of introduction and naturalization of non-native plant species in a wide range of different ecosystems to predict the behaviour of transgenic plants.

Conclusions

Many cultigens that are already transformed or are expected to be transformed can be regarded as less domesticated and will not strongly depend on cultivation. Furthermore there is a growing tendency to introduce traits by genetic engineering that will increase competitiveness and vigour of the transgenic plants in natural and semi-natural ecosystems. Obviously we need powerful instruments - e.g. Models 1 and 2 - to predict the long-term future behaviour of such genetically modified organisms.

References

Kowarik, I., 1992: Einfuhrung und Ausbreitung nichteinheimischer Geholzarten in Berlin und Brandenburg und ihre Folgen fur Flora und Vegetation - Ein Modell fur die Freisetzung gentechnisch veranderter Organismen. - Verh. Bot. Ver. Berlin Brandenburg, Beiheft 3: 1-188.

Raybould, A. F. & Gray, A. J., 1993: Genetically modified crops and hybridization with wild relatives: a UK perspective. - J. Appl. Ecol. 30: 199-219.

Regal, P. J., 1986: Models of Genetically Engineered Organisms and their Ecological Impact. - In: H. A. Mooney & J. A. Drake (Ed.): Ecology of Biological Invasions of North America and Hawaii. Springer-Verlag, New York, Berlin, Heidelberg, London, Paris, Tokyo: pp. 111-129.

Regal, P. J., 1993: The true meaning of 'exotic species' as a model for genetically engineered organisms. - Experientia 49(3): 225-234.

Sharples, F. E., 1982: Spread of organisms with novel genotypes: thoughts from an ecological perspective. - ORNL/TM-8473, Oak Ridge National Laboratory Environmental Sciences Division Publication No. 2040: pp. 1-50.

Sukopp, H. & Sukopp, U., 1993: Ecological long-term effects of cultigens becoming feral and of naturalization of non-native species. - Experientia 49(3): 210-218.

Sukopp, U. & Sukopp, H., 1993: Das Modell der Einfuhrung und Einburgerung nicht einheimischer Arten - Ein Beitrag zur Diskussion uber die Freisetzung gentechnisch veranderter Kulturpflanzen. - Gaia (5): 267-288.

Vries, F. T. de, Meltden, R. van der & Brandenburg, W. A., 1992: Botanical Files - A study of the real chances for spontaneous gene flow from cultivated plants to the wild flora of the Netherlands. - Gorteria Supplement 1: 1-100.

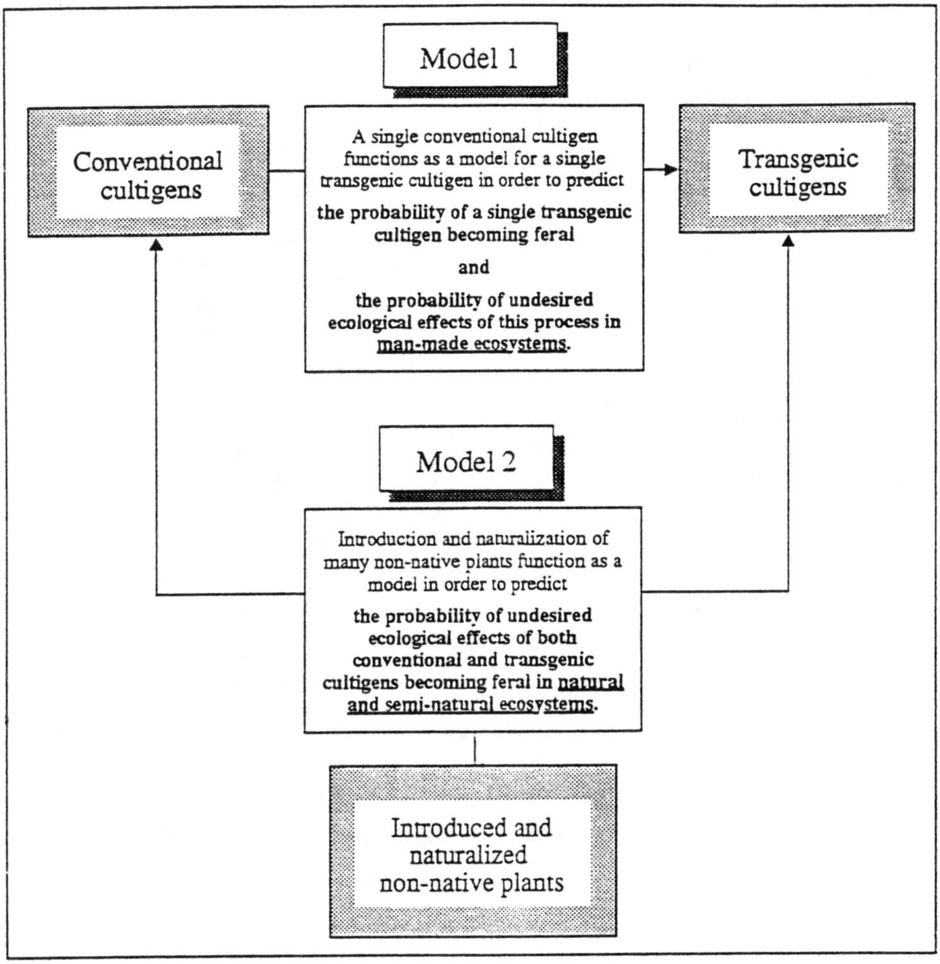

Figure 1: Ecological models of conventional and transgenic cultigens.

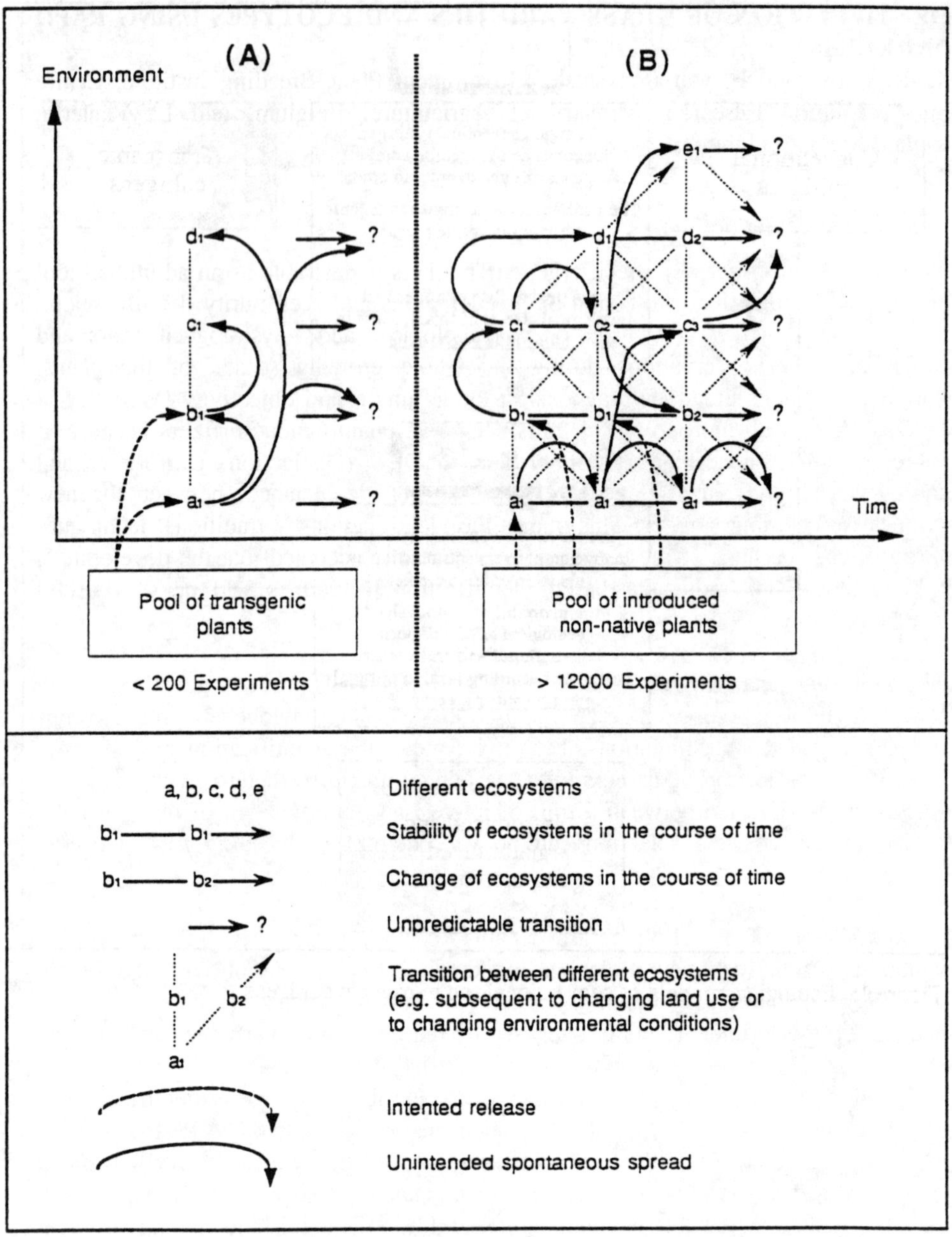

Figure 2: Spatial and temporal range of human knowledge about the ecological behaviour of (A) transgenic plants and (B) introduced non-native plants in different ecosystems (according to KOWARIK 1992).

IDENTIFICATION OF GRASS VARIETIES AND ECOTYPES USING RAPD MARKERS

M. de. Loose and E. van Bockstaele, Government Plant Breeding Institute, Centre for Agricultural Research, Ministry of Agriculture, Belgium, and L. Vauterin, Applied Maths, Belgium

Introduction

The development of variety specific genetic markers is desirable as an additional tool for variety identification, protection of breeder rights and seed purity determination.. Traditional techniques are based on morphological and physiological traits and require extensive observations during the whole growing season of the plants. Moreover, in some situations diagnosis lacks definition and objectivity (Wringley at al. 1987), as not all morphological traits can serve unambiguous markers because of environmental influences. Furthermore due to DUS (Distinction, Uniformity and Stability) rules and due to breeder efforts, the genetic distances between the new varieties is becoming smaller. This makes the identification by traditional techniques more difficult. because of these reasons it is generally accepted that the development of biochemical and molecular based identification techniques will be very useful (Yang and Quiros 1993).

The molecular analysis of varieties and ecotypes will provide geneticists with interesting information, which can be used to evaluate phylogenetic relationships among heterogeneous populations. Moreover, molecular identification of genotypes will allow the selection of parents with the aim to maximize heterosis in crosses. If linkage can be detected between a trait of interest and the presence of the molecular marker, marker assisted selection can be very helpful in breeding for interesting properties..

A modification of the Polymerase Chain Reaction (PCR) has resulted in the availability of relatively new molecular markers: the Random Amplified Polymorphic DNA (RAPD) markers (Williams et al. 1990). Polymorphisms are caused by base changes in the primer binding sites or by chromosome rearrangements in the amplified sequences, resulting in different amplified fragment patterns.
In this paper we evaluate the possibilities of RAPD markers for the identification of rye grass varieties and ecotypes. By computer analysis we compared RAPD patterns from 22 rye grasses from the Government Plant Breeding Institute (RvP) (*Lolium multiforum westerwoldicum* (2), *Lolium multiforum* (6) and *Lolium perenne* (14)). We show that this method is rapid and useful for distinguishing among different varieties and clustering of genotypes in the *Lolium* complex.

Material and methods

The plants were grown in the field. The DNA was prepared by a standard phenol extraction as described by Dellaporte et al. (1983). For the varieties Bellem and Lemundo DNA was prepared from individual plants, as well as from bulked leaf samples harvested on five plants. For the other 20 varieties, DNA preparations were made only from bulked leaf samples from ten individual grass plants.

The RAPD screening was done by using 60 different ten-mer primers purchased from Operon Inc.. The PCR reactions were essentially performed as described by Williams et al. (1990), and the obtained DNA fragments separated by 2% agarose gelelectrophoresis. The picture with the DNA profiles were scanned with a HP deskscan II_p. Finally, these densitometric RAPD scanning patterns were compared via the GELCOMPAR software programm (Applied Maths, Kortrijk, Belgium).

Results

Sixty primers were evaluated in an RAPD analysis on an Italian and a perennial rye-grass genotype. All primers yielded defined fragment patterns but only seven of them gave rise to clearly different profiles. These seven primers were retained for the detailed analysis and comparison of all the different RvP rye grass varieties. For each primer a multiple band profile or fingerprint was produced comprising one to five major bands and a varying number of minor bands. Certain amplified bands appeared to be common to several varieties while others were present in some species but absent in others.

For each variety three different PCR reactions with primers OPB-20, OPC-9 and OPC-15 were loaded on an agarose gel in slots made in series under each other. In this way the three different patterns combine approximately 50 loci. These can be screened and analyzed per lane by the GELCOMPAR software. After scanning the gels, the profiles were normalized based on a molecular marker. We demonstrated that combined patterns are reproducible in repeated RAPD experiments on a given DNA preparation as well as on different DNA preparations of the same plant(s).

The scanning data were compared by the Pearson correlation coefficient. Cluster analysis was performed using UPGMA clustering algorithm (Vauterin and Vauterin, 1992). The resulting dendrogram could reflect the phylogenetic relationship, but it is not necessarily a reflection of the morphological relatedness. The strong relationship between many varieties is expected as most of them are build up of related genetic components. In another experiment we have shown that there are some differences between patterns obtained with DNA isolated from individual plants and the patterns obtained with bulked DNA from the same plants. This is not surprising because each rye grass is a synthetic variety and consists of different related components. Moreover, many factors can influence the equilibrium of which fragments will be amplified in the PCR reaction. By using bulked samples possible polymorphisms in these group can be masked. Nevertheless, the GELCOMPAR software allows to classify all patterns from a given variety in one group.

Conclusion

In this paper we describe a procedure for the identification of rye grasses. By combining multiple RAPD profiles in the GELCOMPAR software analysis, we were able to produce reproducible dendrograms for different varieties of perennial and Italian rye grasses. Therefore, we expect that it will be possible to adapt this strategy for the identification and the clustering of genotypes and ecotypes in other cross pollinating species.

References

Dellaporte, S.L., Wood, J. and Hicks, J.B. (1983). Plant. Mol. Biol. Reporter, 1, 19-21.

Vauterin, L. and Vauterin, P. (1992) Eur. Microbiol., 1, 37-41.

Williams, J.G.K., Kubelik, A.E., Levak, K.J., Rafalski, J.A. and Tinney, S.C. (1990) Nucleic Acids Res. 18, 6531-6535.

Wringley, C.W., Batey, I.L. and Skerritt, J.H. (1987) Seed Sci. Technol. 15, 679-688.

Yang, X. and Quiros, C. (1993) Theor. Appl. Genet. 86, 205-212.

able to produce reproducible fingerprints for different varieties of perennial and [illegible] rye grasses. Therefore, we expect it to be quite possible to adapt this strategy for the identification and the discovery of genotypes and ecotypes in other cross pollinating species.

References

[illegible] S.L., Woo[illegible] and [illegible] (19[illegible]) Plant Mol. Biol. Reporter [illegible]

Vandenb[illegible] and Zabeau [illegible] (1992) [illegible] Mol. Biol. [illegible]

[illegible] Zabeau [illegible] (19[illegible]) [illegible] Nucleic [illegible]

Wen[illegible] [illegible] (198[illegible]) [illegible] Sci. [illegible]

[illegible] (19[illegible]) [illegible]

MEMBERS OF THE ORGANIZING COMMITTEE

Dr. F. Turnowsky
Institut für Mikrobiologie, Universität Graz
Universitätsplatz 2
A- 8010 Graz, Austria

Dr. H.E. Svart
The National Forest and Nature Agency, Ministry of Environment
Haraldsgade 53
DK-2100 Copenhagen, Denmark

Dr. M. Tepfer
Laboratoire de Biologie Cellulaire, Centre National de Recherche Agronomique
Route de St. Cyr.
F-78000 Versailles, France

Dr. L. Fiedler
Bundesministerium für Umwelt, Naturschutz und Reaktorsicherheit
P.O. Box 120629
D-5300 Bonn 1, Germany

Dr. E. Balázs
Agricultural Biotechnology Center
P.O. Box 170
H-2101 Gödöllö, Hungary

Dr. R. Harrington
Mammal Conservation Research, National Parks and Wildlife Service
Newton-mount-kennedy Co Wicklow, Ireland

Prof. dr. P.J. Schembri
Department of Biology, University of Malta
Msda Malta, Malta

Mr. drs. P.J. van der Meer (chair)
Directorate-General for Environment Protection / SVS IPC 655, Ministry of Housing, Physical
Planning and Environment
P.O. Box 30945
NL-2500 GX Den Haag, The Netherlands

Mr. J. Husby
Directorate for Nature Management
Tungasletta 2
7004 Trondheim, Norway

Mrs. E. Barahona / mrs. L. Roda
Direción General de Politica Ambiental, Ministerio de Obras Públicas, Transportes y Medio
Ambiental
Paseo de la Castelana 67
E-28071 Madrid, Spain

Mr. E. Kohli
Division Protection Nature, Office Fédéral de l'Environnement des Forêts et du Paysage
Case Postale
CH-3003 Berne, Switzerland

Dr. H. Marquard
Biotechnology Unit, Regulatory Section, Department of the Environment
Room B361, Romney House
43 Marsham Street
London SW1P 3PY, United Kingdom

RAPPORTEURS

Prof. dr. S. Tolin
Department of Plant Pathology, Physiology & Weed Science, Virginai Polytechnic Institute & State
University
Blacksburg
Virginia 24061-0330, United States of America

Dr. M. Tepfer
Laboratoire de Biologie Cellulaire, Centre National de Recherche Agronomique
Route de St. Cyr.
F-78000 Versailles, France

Prof. dr. J. Beringer
Department of Botany, School of Biological Sciences, University of Bristol
Woodland Road
Bristol BS8 1UG, United Kingdom

Dr. W. Moens
Institute of Hygiene and Epidemiology, Federal Ministry of Health and Environment
Rue Juliette Wytsman 14
B-1050 Brussels, Belgium

Dr. C. Pla
Laboratory of Ichthyology (Genetics), University of Girona
Plaza Hospital 6
E-17071 Girona, Spain

Dr. R. Harrington (General Rapporteur)
Mammal Conservation Research, National Parks and Wildlife Service
Newton-mount-kennedy Co Wicklow, Ireland

SPEAKERS

Prof. dr. J. Schell
Max Planck Institut für Züchtungsforschung
Carl von Linné Weg 10
D-50829 Köln 30, Germany

Dr. M. Crawley
Centre for Population Biology, Imperial College
Silkwood Park Ascot
Berkshire SL5 7PW, United Kingdom

Dr. M. Peferoen
Crop Protection, Plant Genetic Systems N.V.
Jozef Plateaustraat 22
B-9000 Gent, Belgium

Dr. R. Hengeveld
Institute for Forest & Nature Research
P.O. Box 23
NL-6700 AA Wageningen, The Netherlands

Dr. E. Balázs
Agricultural Biotechnology Center
P.O. Box 170
H-2101 Gödöllö, Hungary

Dr. F. García-Arenal
Laboratorio de Patologia Vegetal, ETSI Agronomos
Cuidad Universitaria
E-28040 Madrid, Spain

Dr. J.M. Thresh
Natural Resources Institute
Central Avenue
Chatham Maritime
Kent ME4 4TB, United Kingdom

Dr. R. Possee
Institute of Virology and Environmental Microbiology
Mansfield Road
Oxford OX1 3SR, United Kingdom

Prof. dr. J.C. Zadoks
Department of Phytopathology, Agricultural University Wageningen
P.O. Box 8025
6700 EE Wageningen, The Netherlands

Prof. P.P. Pastoret
Service de Virologie-Immunologie, Faculté de Médecine Vétérinaire, Université de Liège
Institut de Chimie (B6) - Local R80
B-4000 Sart Tilman
Liège, Belgium

Dr. T. Kimman
Central Veterinary Institute
P.O. Box 365
8200 AJ Lelystad, The Netherlands

Dr. D. Chourrout
Laboratoire de Génétique des Poissons, Institut National de Recherche Agrnomique
Domaine de Vilvert
F-78352 Jouy-en-Josas CEDEX, France

Dr. K. Hindar
Norwegian Institute for Nature Research
Tungasletta 2
7004 Trondheim, Norway

OBSERVERS

Mr. Y. Ando
Biotechnology Unit, Direcorate for Science, Technology and Industry, Organisation for Economic
Co-operation and Development
2 Rue André Pascal
F-75775 Paris CEDEX, France

Mr. M. Cantley
Biotechnology Unit, Direcorate for Science, Technology and Industry, Organisation for Economic
Co-operation and Development
2 Rue André Pascal
F-75775 Paris CEDEX, France

Mrs. L. Zannoni
Environment Directorate, Organisation for Economic Co-operation and Development
2 Rue André Pascal
F-75775 Paris CEDEX, France

Mr. P. Kearns
Environment Directorate, Organisation for Economic Co-operation and Development
2 Rue André Pascal
F-75775 Paris CEDEX, France

Dr. M. Lex
DG XII, Commission of the European Communities
200 Rue de la Loi
B-1049 Brussels, Belgium

Dr. H. Pedersen
DG XI, Commission of the European Communities
200 Rue de la Loi
B-1049 Brussels, Belgium

Mrs. J. Tachmintzis
DG XI, Commission of the European Communities
200 Rue de la Loi
B-1049 Brussels, Belgium

Dr. V. Villalobos
United Nations Food and Agriculture Organisation
Via delle Terme di Caracalla
I-00100 Rome, Italy

SECRETARIAT

Mr. F. Albanese
Director of Environment and Local Authorities, Council of Europe
P.O. Box 431 R6
F-67006 Strasbourg CEDEX, France

Mr. J.P. Ribaut
Head of the Environment Conservation and Management Division, Council of Europe
P.O. Box 431 R6
F-67006 Strasbourg CEDEX, France

Mr. H. Hacourt
Principal Administrator, Environment Conservation and Management Division, Council of Europe
P.O. Box 431 R6
F-67006 Strasbourg CEDEX, France

Mr. P. Schenkelaars
MEBO Environmental Consultancy
P.O. Box 38
2250 AA Voorschoten, The Netherlands

Mr. R. van der Graaf
MEBO Environmental Consultancy
P.O. Box 38
2250 AA Voorschoten, The Netherlands

PARTICIPANTS

AUSTRIA

Mr. H. Gaugitsch
Federal Environmental Agency
Spittelauer Lände 5
A-1120 Vienna, Austria

Dr. M. Laimer da Câmara Machado
Institut für Angewandte Mikrobiologie, Universität für Bodenkultur
Nussdorfer Lände 11
A-1190 Vienna, Austria

BELGIUM

Mrs. L. Bullard
The Green Group of the European Parliament
Mon-316
Rue Belliard
B-1047 Brussels, Belgium

Mr. E. van Bockstaele
R.V.P. Government Plant Breeding Institute
Burgemeester van Gansberghelaan 109
B-9820 Merelbeke, Belgium

CANADA

Dr. J. Hollebone
Biotech Strategies and Co-ordination Officer
Room 495, Sir John Carling bld.
Agriculture & Agrifood Canada
Ottawa K1A OC6, Canada

DENMARK

Mr. R. Bagger Jørgensen
Plant Biology Section, Risø National Laboratory
P.O. Box 49
DK-4000 Roskilde, Denmark

Mr. C. Frier
Danish Environmental Protection Agency
Strandgade 29
DK-1401 Copenhagen K, Denmark

Mr. J.G. Højland
The National Forest & Nature Agenccy
Haraldsgade 53
DK-2100 Copenhagen Ø, Denmark

Mr. G. Kjellson
Department of Terrestrial Ecology, National Environmental Research Institute
P.O. Box 314
DK-8600 Silkeborg, Denmark

Mr. L. Landbo
Risø National Laboratory
P.O. Box 49
DK-4000 Roskilde, Denmark

Mrs. M. Philipp
Department of Plant Ecology, University of Copenhagen
Øster Farimagsgade 2D
DK-1353 Copenhagen, Denmark

FRANCE

Mr. J. Arnould
Laboratoire d'Evolution et Systématiques Végétales, Université de Paris XI - CNRS
Université de Paris XI, Bât 362
F-91405 Orsay CEDEX, France

Mrs. P. Barbier
Institut National de Recherche Agronomique
28 Rue d' Herrlisheim
F-68021 Colmar, France

Mrs. S. Ben Tahar
Biocem, groupe Limagrain
24 Avenue des Landais
F-63170 Aubière, France

Mr. Y. Chupeau
Labo Biochem, Institut National de Recherche Agronomique
F-78026 Versailles CEDEX, France

Mr. R. Delon
Seita, Institut du Tabac
Domaine de la Tour
F-24100 Bergerac, France

Mr. M. Ravelonandro
Chargé de Recherche à la Station de Pathologie Végétale, Institut National de Recherche Agro-
nomique, Centre de Recherches de Bordeaux
P.O. Box 81
F-33883 Villenave d'Ornon CEDEX, France

Mr. J. Marrou
Ministère de l'Agriculture et de la Pêche, DPE
3 Rue Barbet de Jouy
F-75007 Paris, France

Mr. J.-J. Leguay
Elf-Sanofi
8 Rue Chirstophe Colomb
F-75008 Paris, France

Mr. Ch. Putz
Institut National de Recherche Agronomique
28 Rue d'Herrlisheim
F-68021 Colmar CEDEX, France

Mr. J.H. San
Seita, Institut du Tabac
Domaine de la Tour
F-24100 Bergerac, France

Mr. B. Walter
Institut National de Recherche Agronomique
28 Rue d'Herrlisheim
F-68021 Colmar CEDEX, France

GERMANY

Dr. D. Bartsch
Institute of Biology (Ecology), Technical University Aachen
Worringerweg 1
D-52074 Aachen, Germany

Prof. dr. W. Körting
Fish Diseases Research Unit, School of Veterinary Medicine
Bünteweg 17
D-30559 Hannover, Germany

Dr. F. Laplace
Federal Ministry of Science & Technology
Heinemannstraße 2
D-5300 Bonn 2, Germany

Mrs. I. Nöh
Genetechnology Group, Umweltbundesamt
Bismarckplatz 1
D-14191 Berlin 33, Germany

Mr. U. Sukopp
Institut für Ökologie, Technische Universität Berlin
Schmidt-Ott-Straße 1
D-12165 Berlin 62, Germany

Mrs. B. Weber
Öko Institut e.v., Institut für Angewandte Ökologie
Binzengrün 34a
D-7800 Freiburg, Germany

HUNGARY

Dr. R. Vassányi
Institut Central de Recherche Alimentaire
P.O. Box 393
H-1536 Budapest, Hungary

THE NETHERLANDS

Dr. H. Bergmans
Provisional Committee on Genetic Modification
P.O. Box 80.022
NL-3508 TA Utrecht, The Netherlands

Prof. dr. P.G. de Haan
Prof. Ritzema Boslaan 47
NL-3571 CM Utrecht, The Netherlands

Dr. H. Schellekens
Provisional Committee on Genetic Modification
Tinbergenpad 6
NL-2912 Niewerkerk a/d Yssel, The Netherlands

Dr. G.E. de Vries
ProBio Products
Meerweg 6
NL-9625 PJ Overschild, The Netherlands

NORWAY

Mrs. S. Foldal
Norwegian Biotechnology Advisory Board
P.O. Box 8027 Dep
N-0030 Oslo, Norway

Dr. B.J. Tomerås
Norwegian Institute for Nature Research
Tungasletta 2
N-7005 Trondheim, Norway

Mr. S. Mykelbust
Directorate for Nature Research
Tungasletta 2
N-7005 Trondheim, Norway

PORTUGAL

Mrs. M.H. Ribeiro Silva Farrall
Department of Environmental Sciences and Engineering (DCEA), New University of Lisbon
Quinta da Torre
P-2825 Monte da Caparica, Portugal

RUSSIA

Prof. M. Kirpichnikov
Department for Sciences of Life and Biotechnology, Ministry for Science and Technology Policy
of the Russian Federation
Twerskaja St. 11
103905 Moscow, Russia

Dr. K.G. Skryabin
Center of Bioengineering of the Russian Academy of Science
Vavilov St 34/5
117334 Moscow, Russia

Dr. N.G. Stepanova
Department for Sciences of Life and Biotechnology, Ministry for Science and Technology Policy
of the Russian Federation
Twerskaja St. 11
103905 Moscow, Russia

SLOVENIA REPUBLIC

Prof. dr. M. Grabnar
Biology Department, Biotechnology Faculty, University of Ljubljana
Jamnikarjeva 101
61000 Ljubljana, Slovenia Republic

SWEDEN

Mr. L. Espeby
Swedish Environmental Protection Agency
S-171 85 Solna, Sweden

Mrs. B. Hedlund
Swedish Environmental Protection Agency
S-171 85 Solna, Sweden

Mr. P. Tenning
Biosafety Officer, Hilleshøg AB / Sandoz Seed
P.O. Box 302
S-26123 Wandsterona, Sweden

SWITZERLAND

Dr. P. Ahl Goy
Ciba-Geigy SA, R-1001A.1.14
CH-4002 Basel, Switzerland

Dr. D. Amman
Büro für Umweltchemie
Zentralstraße 43
CH-8003 Zürich, Switzerland

Dr. G.F. Collet
Physiologie Végétale, Station Fédérale de Recherches Agronomiques de Changins
CH-1260 Nyon, Switzerland

Dr. K. Dorsch-Häsler
Swiss Committee for Biological Safety
Apfelbaumstraße 43
CH-8050 Zürich, Switzerland

Mrs. Y. Jacot
Laboratoire de Phanéiogamie, Université de Neuchatel
Chantemerle 22
CH-2007 Neuchatel, Switzerland

Mrs. J. Keller
Université de Berne
Altenbergrain 21
CH-3013 Berne, Switzerland

Mr. F. Pythoud
BUWAL/Division Protection Nature, Office Fédéral de l'Environnement des Forêts et du Paysage
Case Postale
CH-3003 Berne, Switzerland

UKRAINE

Mrs. M. Stolina
Institute of Molecular Biology & Genetics, Ukraine Academy of Science
Zabolotny St 150
252627 Kiev, Ukraine

UNITED KINGDOM

Mrs. K. Scott
Rowett Research Institute
Greenburn Road, Bucksburn
Aberdeen AB2 9SB, United Kingdom

Dr. D. Bosworth
Health & Safety Executive
Room 417, Magdalen House, Stanley Precinct
Bootle L20 3Q2 (Merseyside), United Kingdom

Dr. A. Gray
Institute of Terrestrial Ecology, Furzebrook Research Station
Wareham BH20 5AS (Dorset), United Kingdom

Mr. R.K. Hay
Scottish Agricultural Science Agency, East Graigs
Edinburgh EH12 8NJ, United Kingdom

Mrs. J. Hill
The Green Alliance
Wellington Street 49
London WC2E 7BN, United Kingdom

Mr. R. Hull
Virus Research Department, John Innes Institute
Conet Lane
Norwich NR4 7UH, United Kingdom

Mr. N. Poole
Zeneca Seeds
Jealott's Hill
Bracknell (Berkshire), United Kingdom

Dr. G.M. Schofield
Unilever Research
Colworth House
Sharnbrook MK44 1LQ (Bedfordshire), United Kingdom

Sales agents for publications of the Council of Europe
Agents de vente des publications du Conseil de l'Europe

AUSTRALIA/AUSTRALIE
Hunter Publications, 58A, Gipps Street
AUS-3066 COLLINGWOOD, Victoria

AUSTRIA/AUTRICHE
Gerold und Co., Graben 31
A-1011 WIEN 1

BELGIUM/BELGIQUE
La Librairie européenne SA
50, avenue A. Jonnart
B-1200 BRUXELLES 20

Jean de Lannoy
202, avenue du Roi
B-1060 BRUXELLES

CANADA
Renouf Publishing Company Limited
1294 Algoma Road
CDN-OTTAWA ONT K1B 3W8

CYPRUS/CHYPRE
MAM
The House of the Cyprus Book
PO Box 1722, CY-NICOSIA

DENMARK/DANEMARK
Munksgaard
Book and Subscription Service
PO Box 2148
DK-1016 KØBENHAVN K

FINLAND/FINLANDE
Akateeminen Kirjakauppa
Keskuskatu 1, PO Box 218
SF-00381 HELSINKI

GERMANY/ALLEMAGNE
UNO Verlag
Poppelsdorfer Allee 55
D-53115 BONN

GREECE/GRÈCE
Librairie Kauffmann
Mavrokordatou 9, GR-ATHINAI 106 78

IRELAND/IRLANDE
Government Stationery Office
Publications Section
4-5 Harcourt Road, IRL-DUBLIN 2

ISRAEL/ISRAËL
ROY International
PO Box 13056
IL-61130 TEL AVIV

ITALY/ITALIE
Libreria Commissionaria Sansoni
Via Duca di Calabria, 1/1
Casella Postale 552, I-50125 FIRENZE

LUXEMBOURG
Librairie Bourbon
(Imprimerie Saint-Paul)
11, rue Bourbon
L-1249 LUXEMBOURG

NETHERLANDS/PAYS-BAS
InOr-publikaties, PO Box 202
NL-7480 AE HAAKSBERGEN

NORWAY/NORVÈGE
Akademika, A/S Universitetsbokhandel
PO Box 84, Blindern
N-0314 OSLO

PORTUGAL
Livraria Portugal, Rua do Carmo, 70
P-1200 LISBOA

SPAIN/ ESPAGNE
Mundi-Prensa Libros SA
Castelló 37, E-28001 MADRID

Llibreria de la Generalitat
Rambla dels Estudis, 118
E-08002 BARCELONA

Llibreria de la Generalitat de Catalunya
Gran Via Jaume I, 38, E-17001 GIRONA

SWEDEN/SUÈDE
Aktiebolaget CE Fritzes
Regeringsgatan 12, Box 163 56
S-10327 STOCKHOLM

SWITZERLAND/SUISSE
Buchhandlung Heinimann & Co.
Kirchgasse 17, CH-8001 ZÜRICH

BERSY
Route du Manège 60
CP 4040
CH-1950 SION 4

TURKEY/TURQUIE
Yab-Yay Yayimcilik Sanayi Dagitim Tic Ltd
Barbaros Bulvari 61 Kat 3 Daire 3
Besiktas, TR-ISTANBUL

UNITED KINGDOM/ROYAUME-UNI
HMSO, Agency Section
51 Nine Elms Lane
GB-LONDON SW8 5DR

UNITED STATES and CANADA/
ÉTATS-UNIS et CANADA
Manhattan Publishing Company
468 Albany Post Road
PO Box 850
CROTON-ON-HUDSON, NY 10520

STRASBOURG
Librairie internationale Kléber
1, rue des Francs-Bourgeois
F-67000 STRASBOURG

Librairie des Facultés
2-12, rue de Rome
F-67000 STRASBOURG

Librairie Kléber
Palais de l'Europe
F-67075 STRASBOURG Cedex

Council of Europe Press/Les éditions du Conseil de l'Europe
Council of Europe/Conseil de l'Europe
F-67075 Strasbourg Cedex